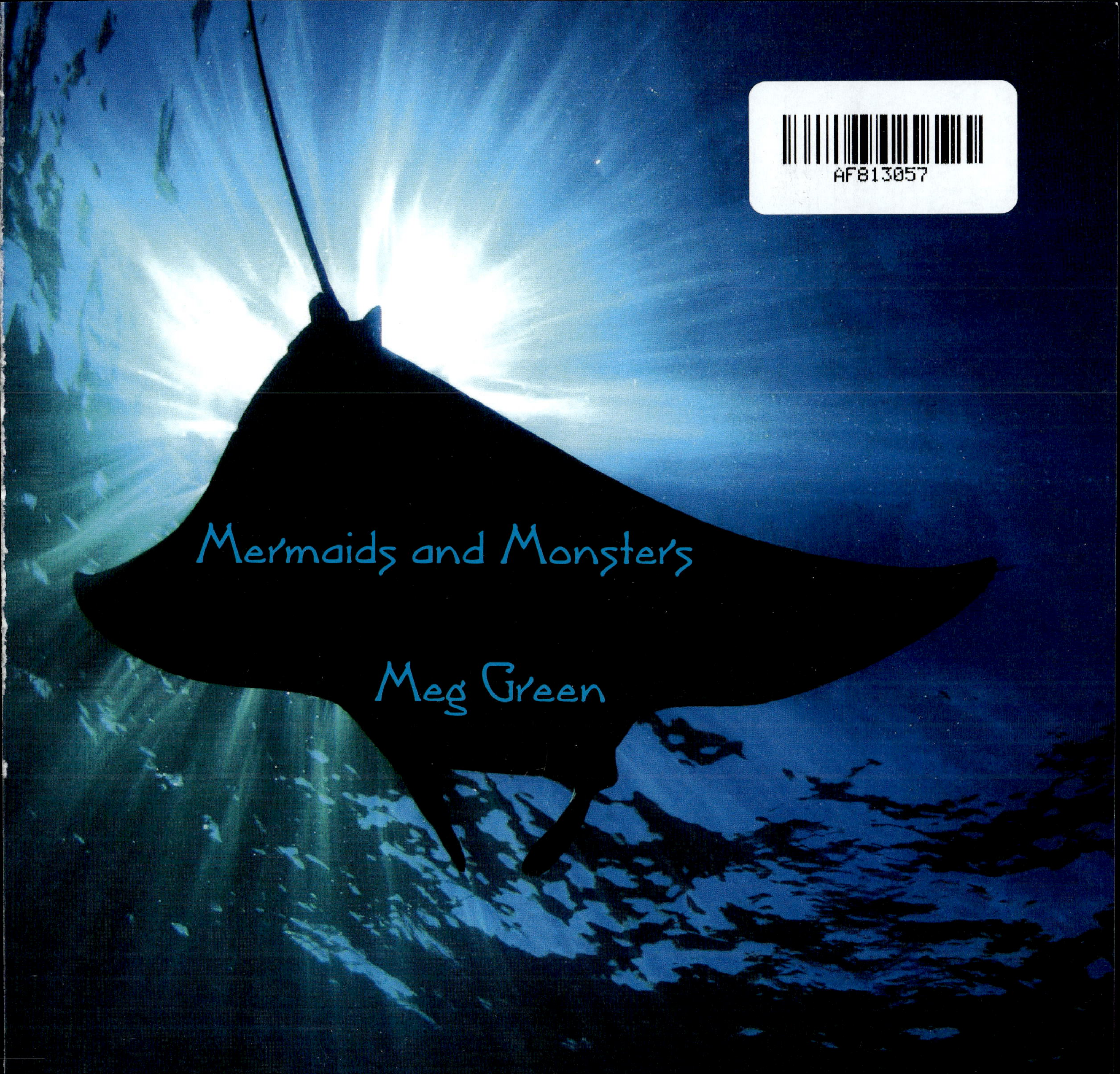

Mermaids and Monsters

Meg Green

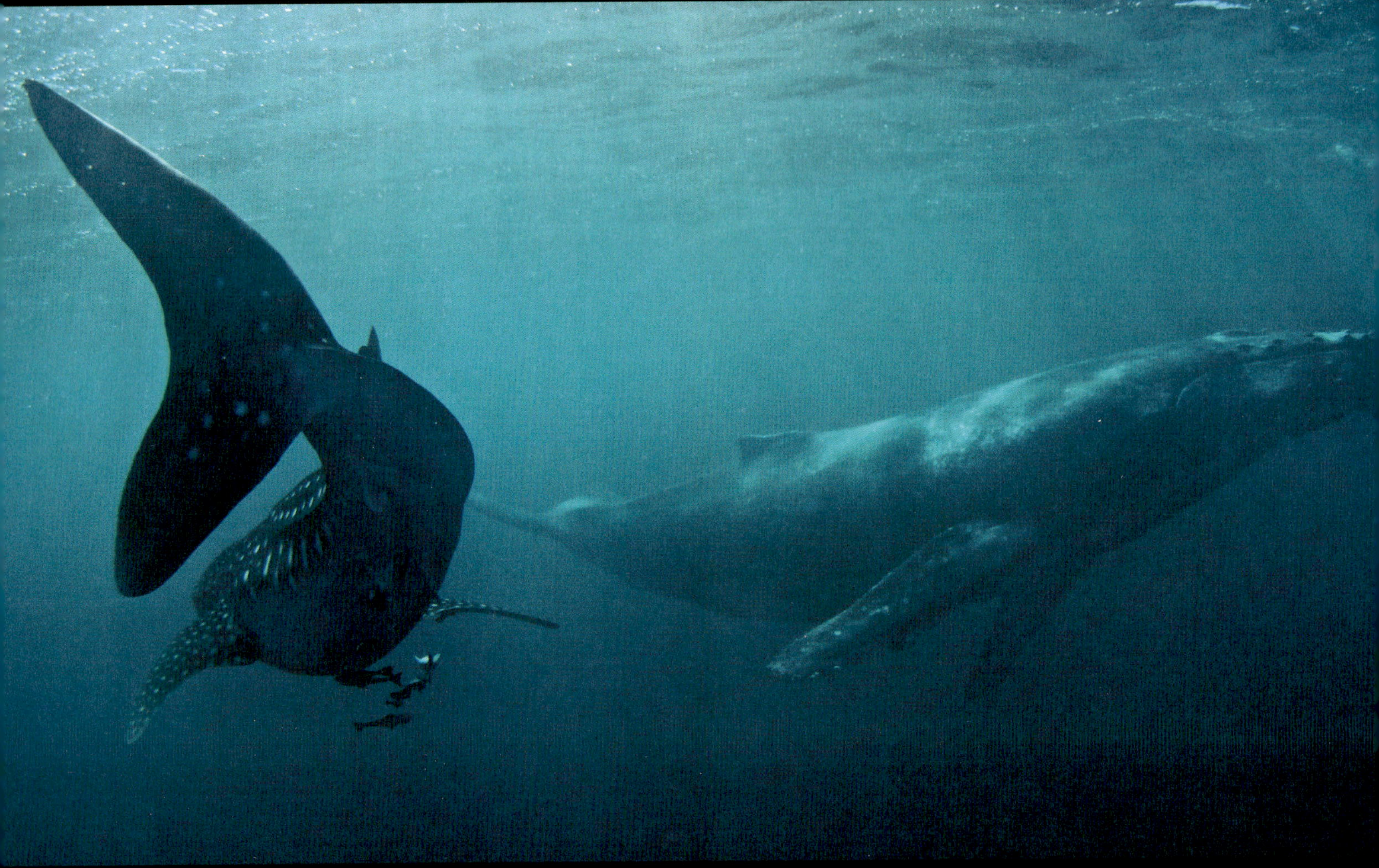

"The difference between fiction and reality?
Fiction has to make sense."
- Tom Clancey

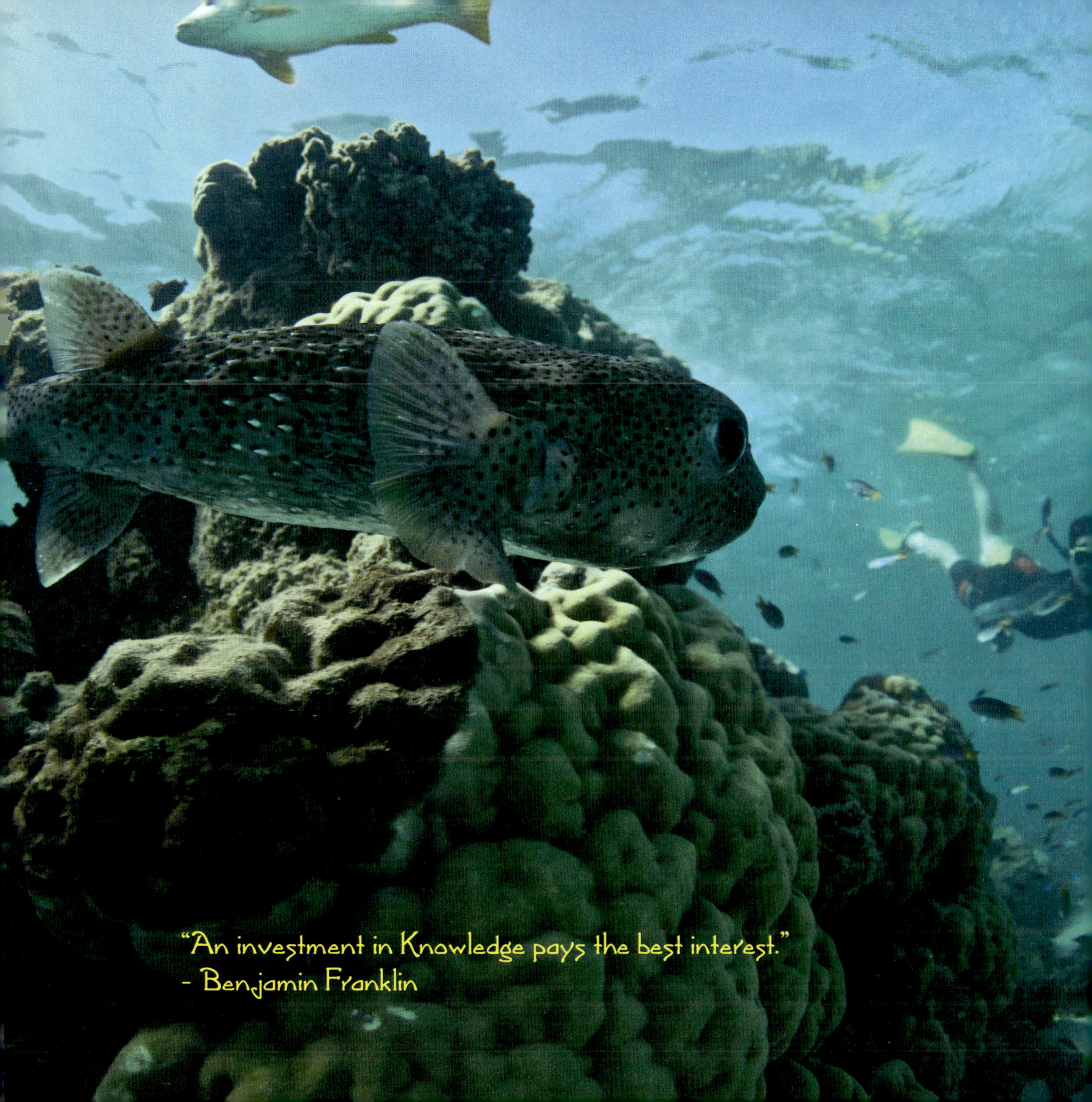

"An investment in Knowledge pays the best interest."
- Benjamin Franklin

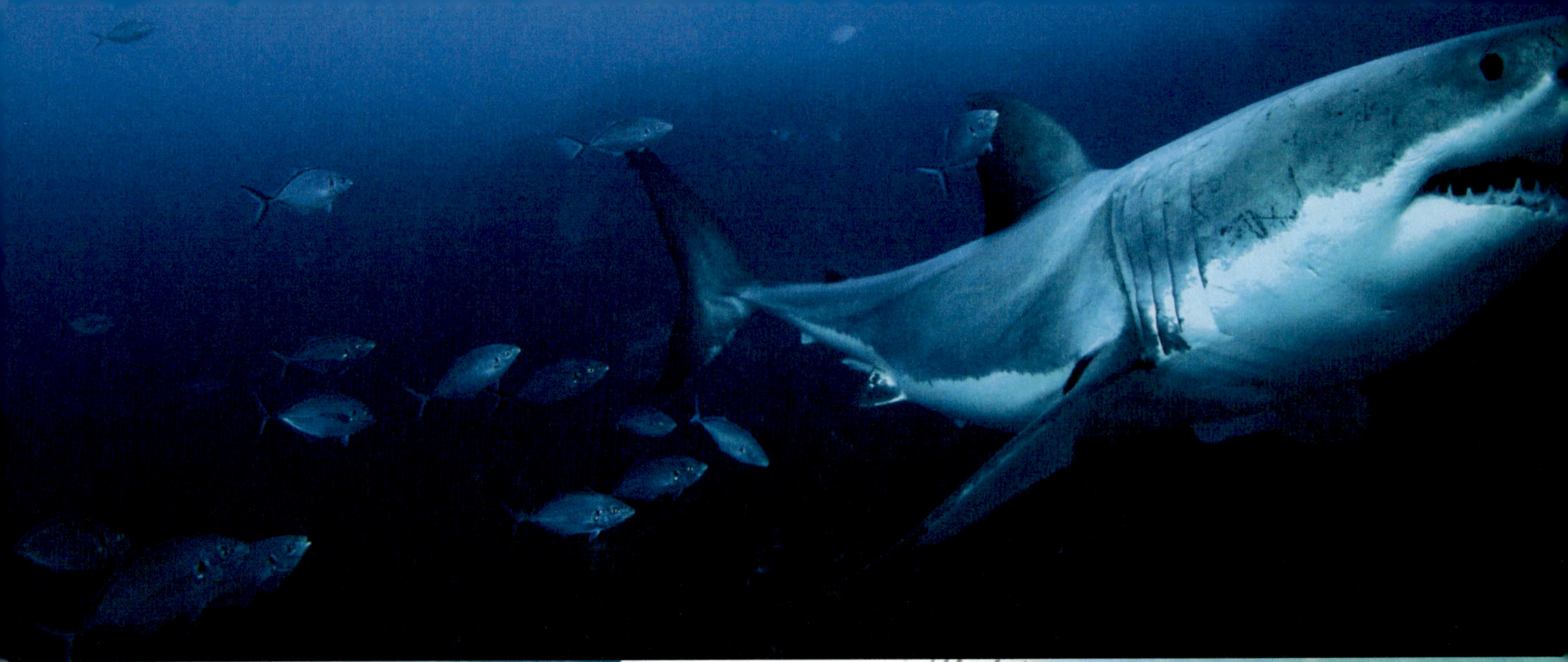
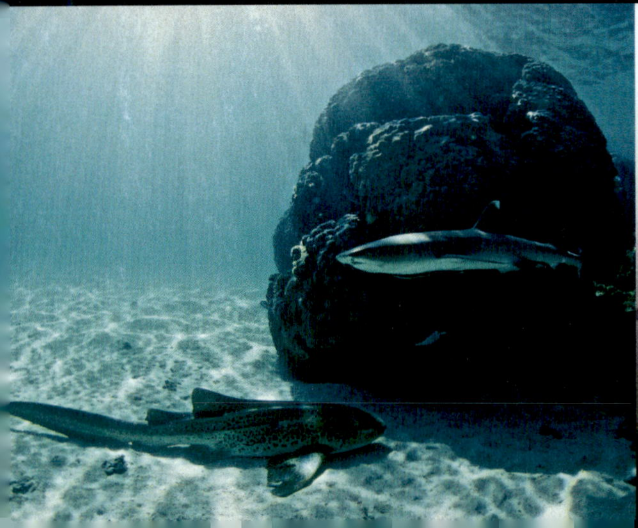
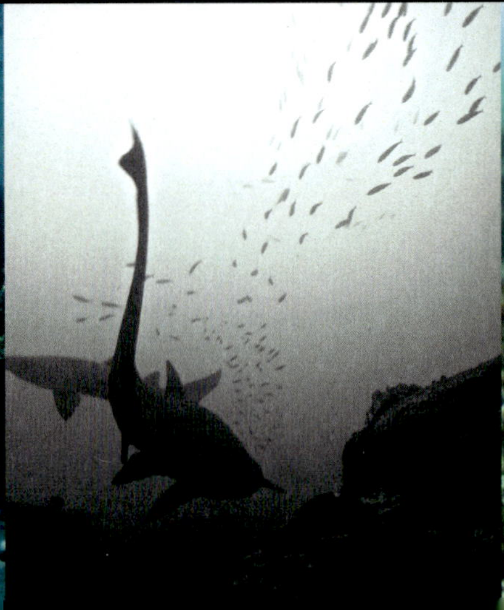
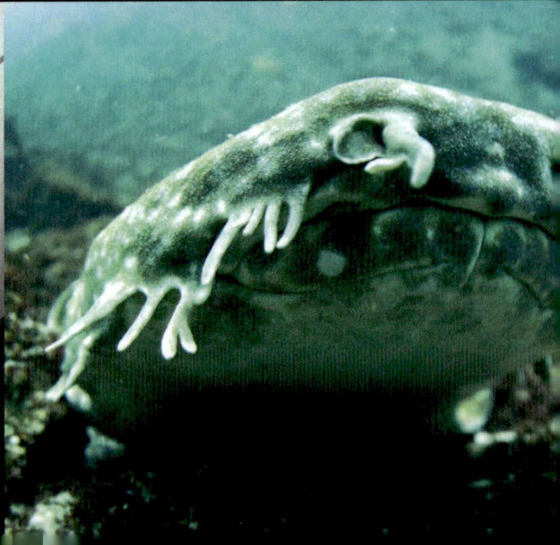

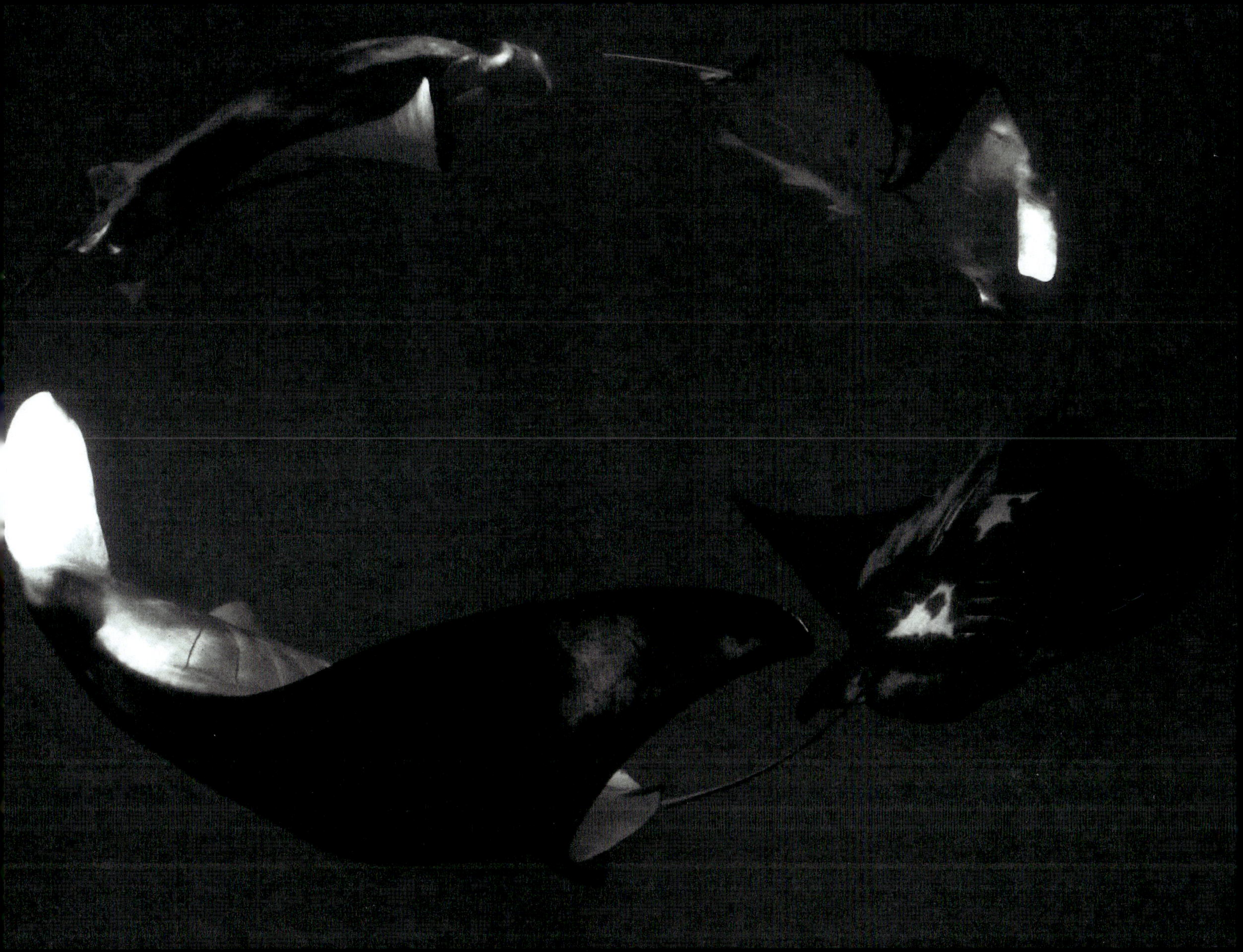

"Human being, vegetable, or cosmic dust – we all dance to a mysterious tune, intoned in the distance by an invisible piper."
– Albert Einstein

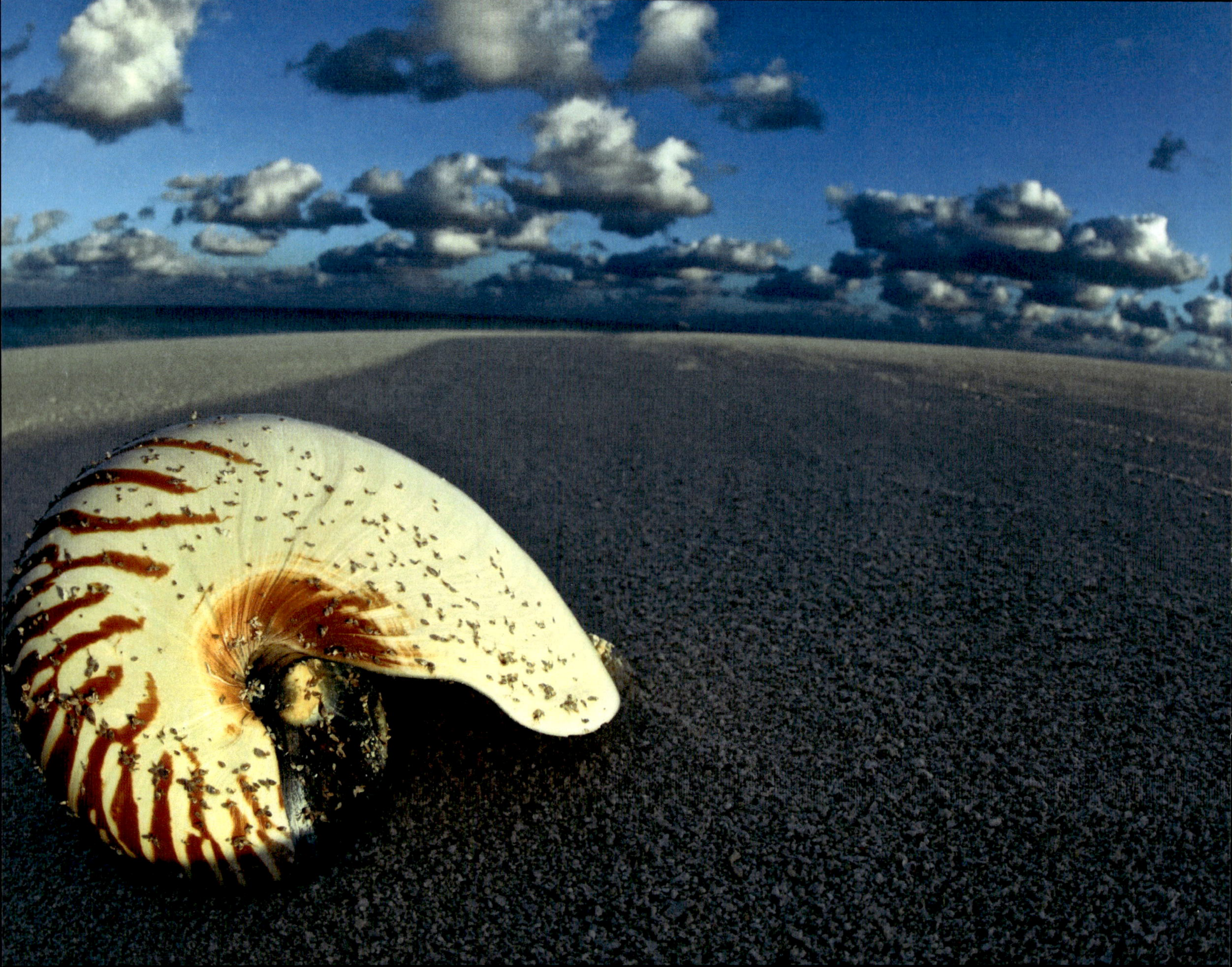

"The larger the island of knowledge, the longer the shore line of wonder."
- Ralph W. Sockman

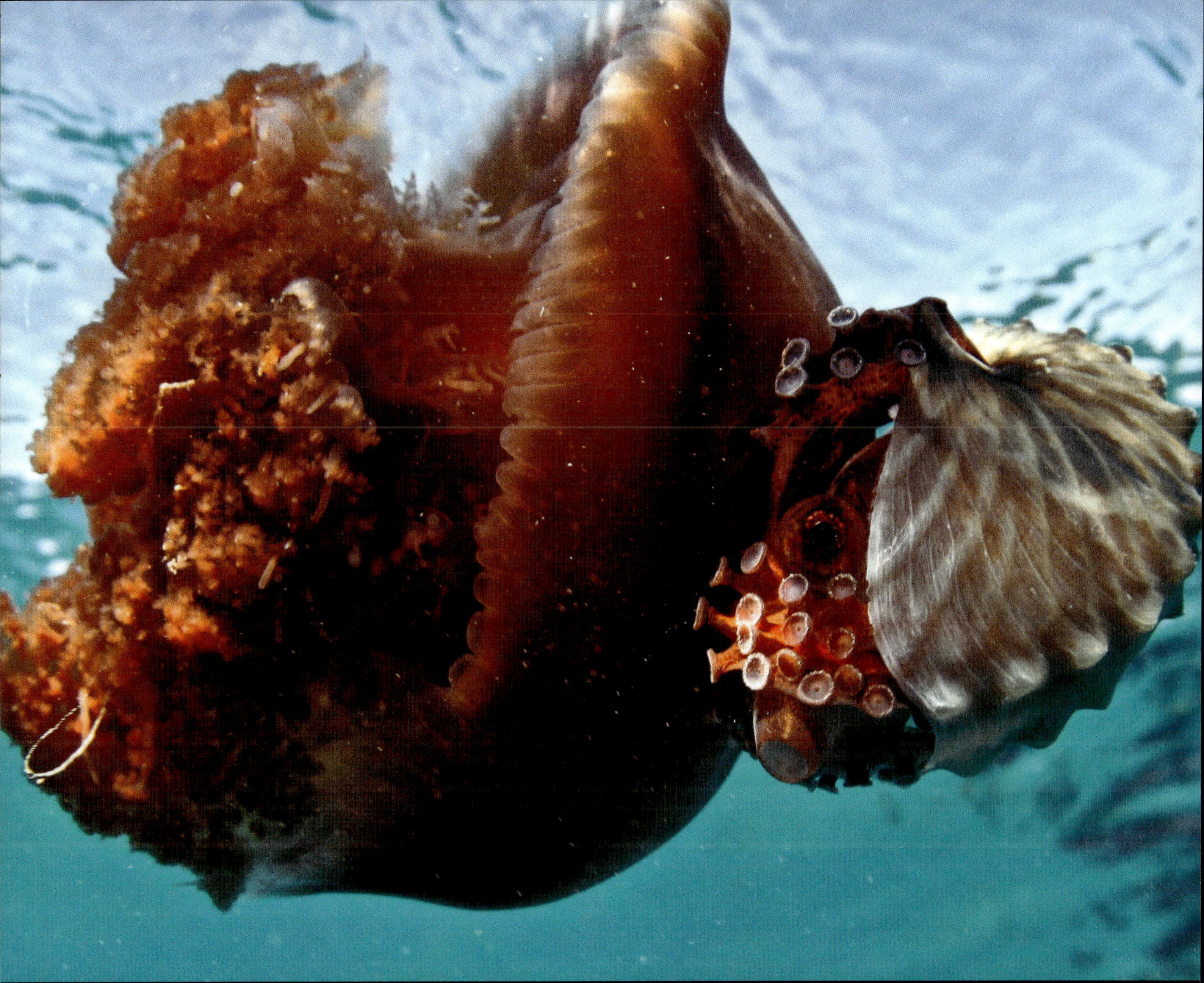

"This world is but a canvas to our imagination."
- Henry David Thoreau

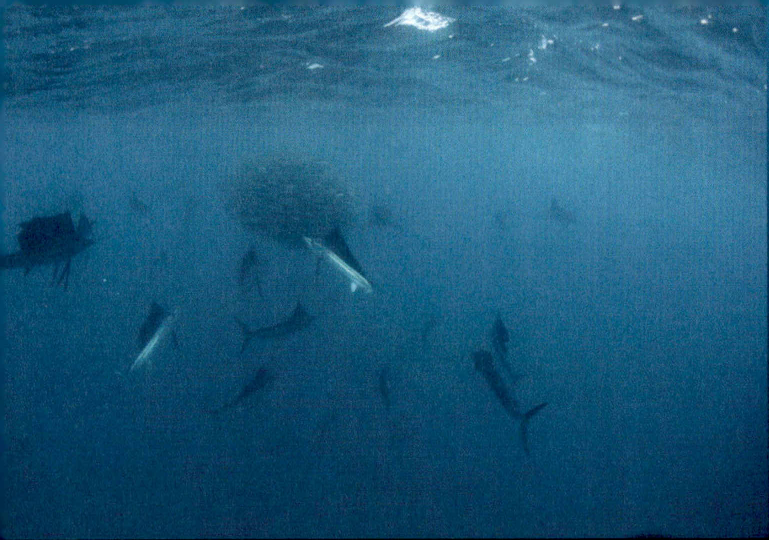

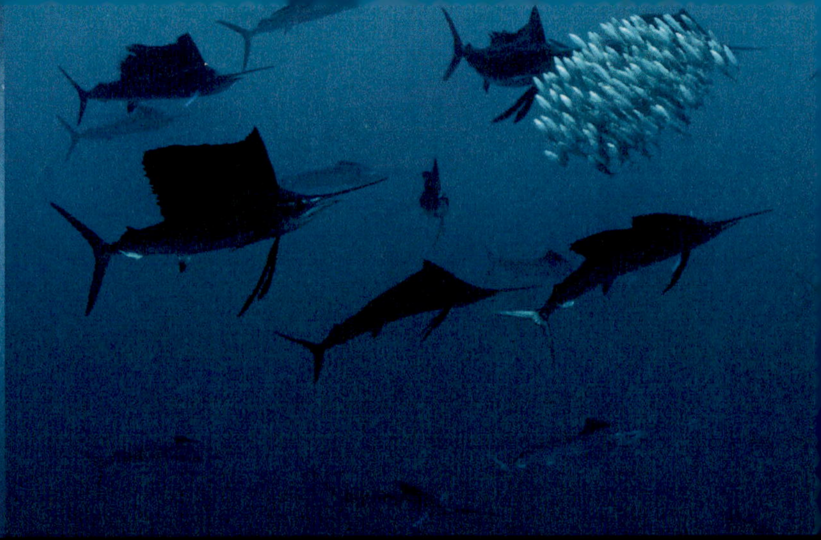

"Coming together is a beginning;
keeping together is progress;
working together is success."
- Henry Ford

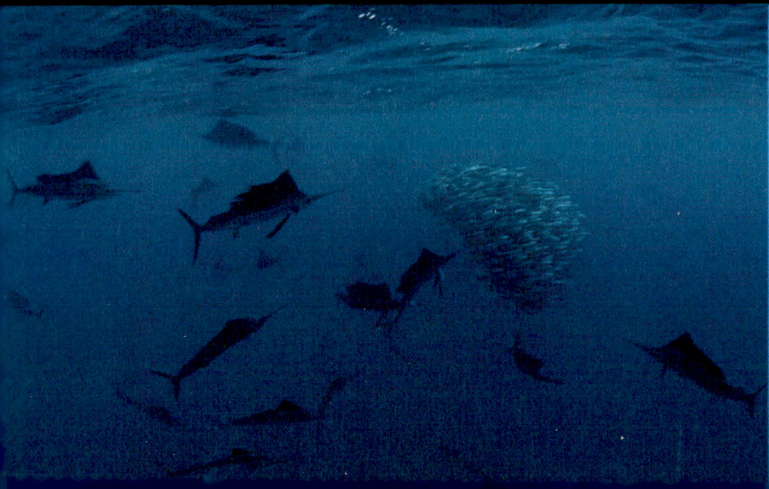

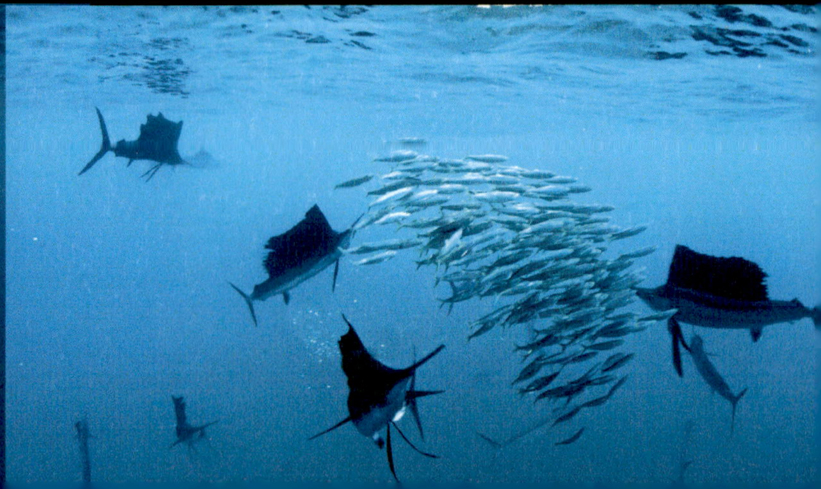

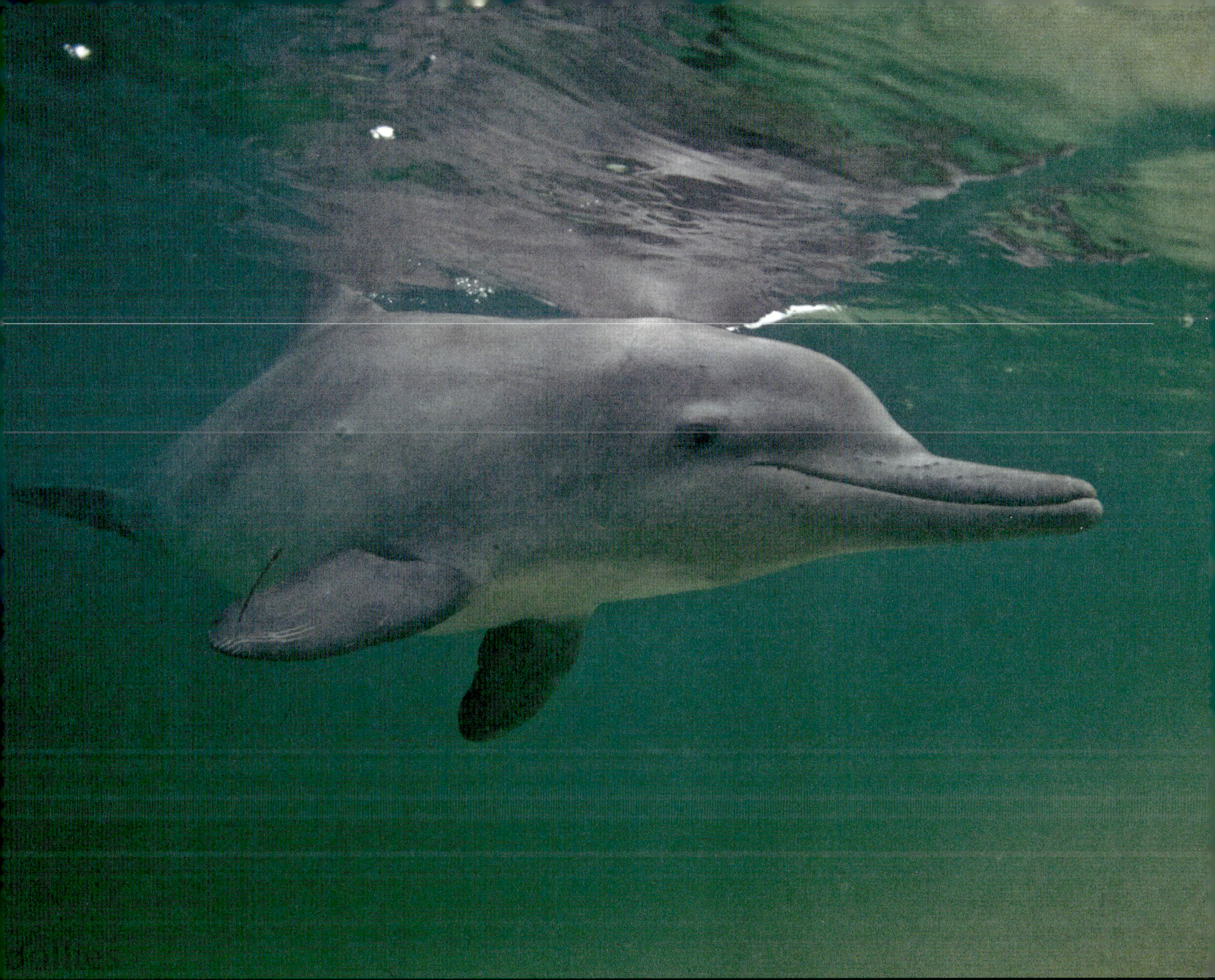

"Always laugh when you can. It is cheap medicine."
- Lord Byron

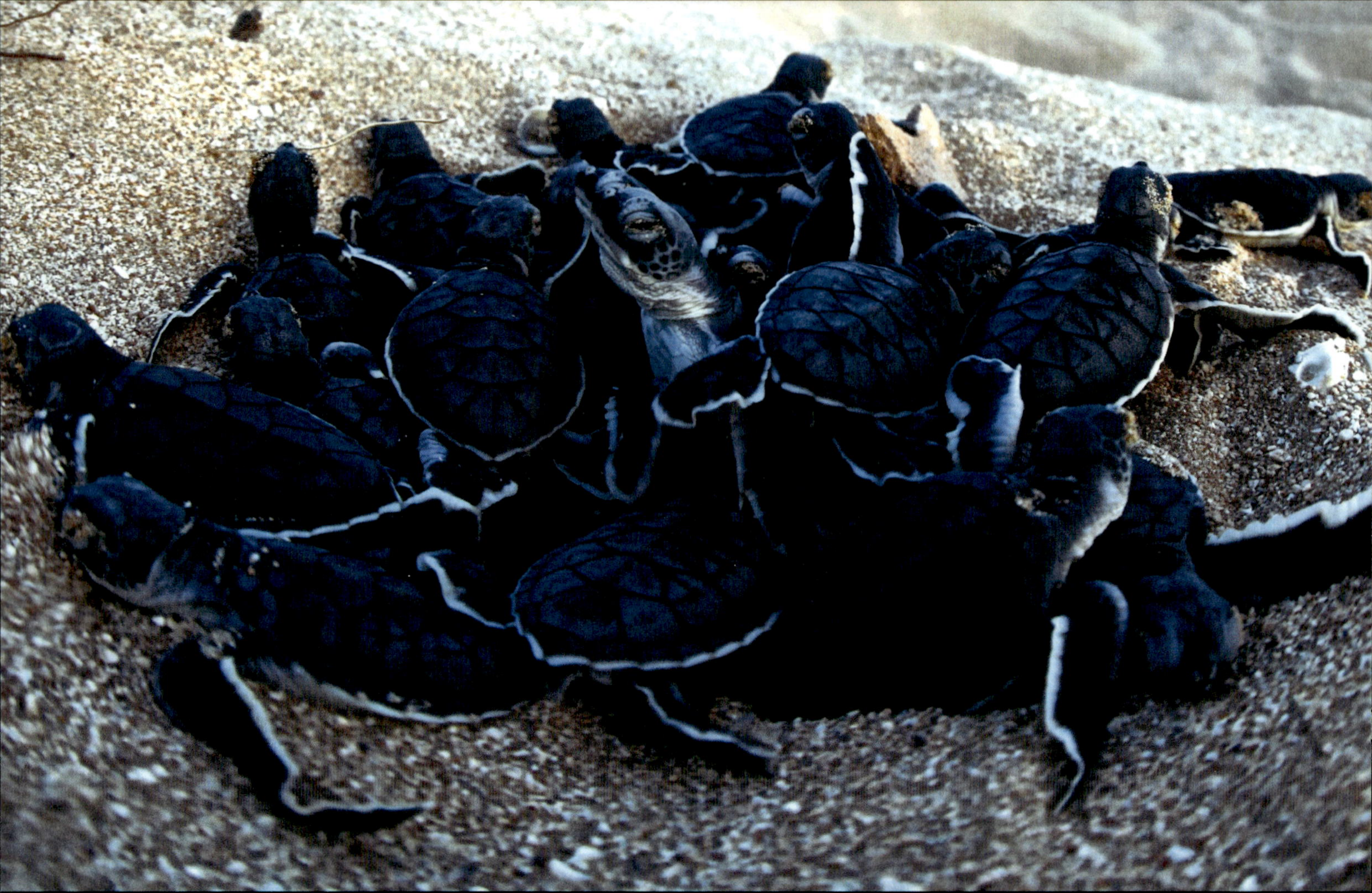

"Strength does not come from physical capacity.
It comes from an indomitable will."

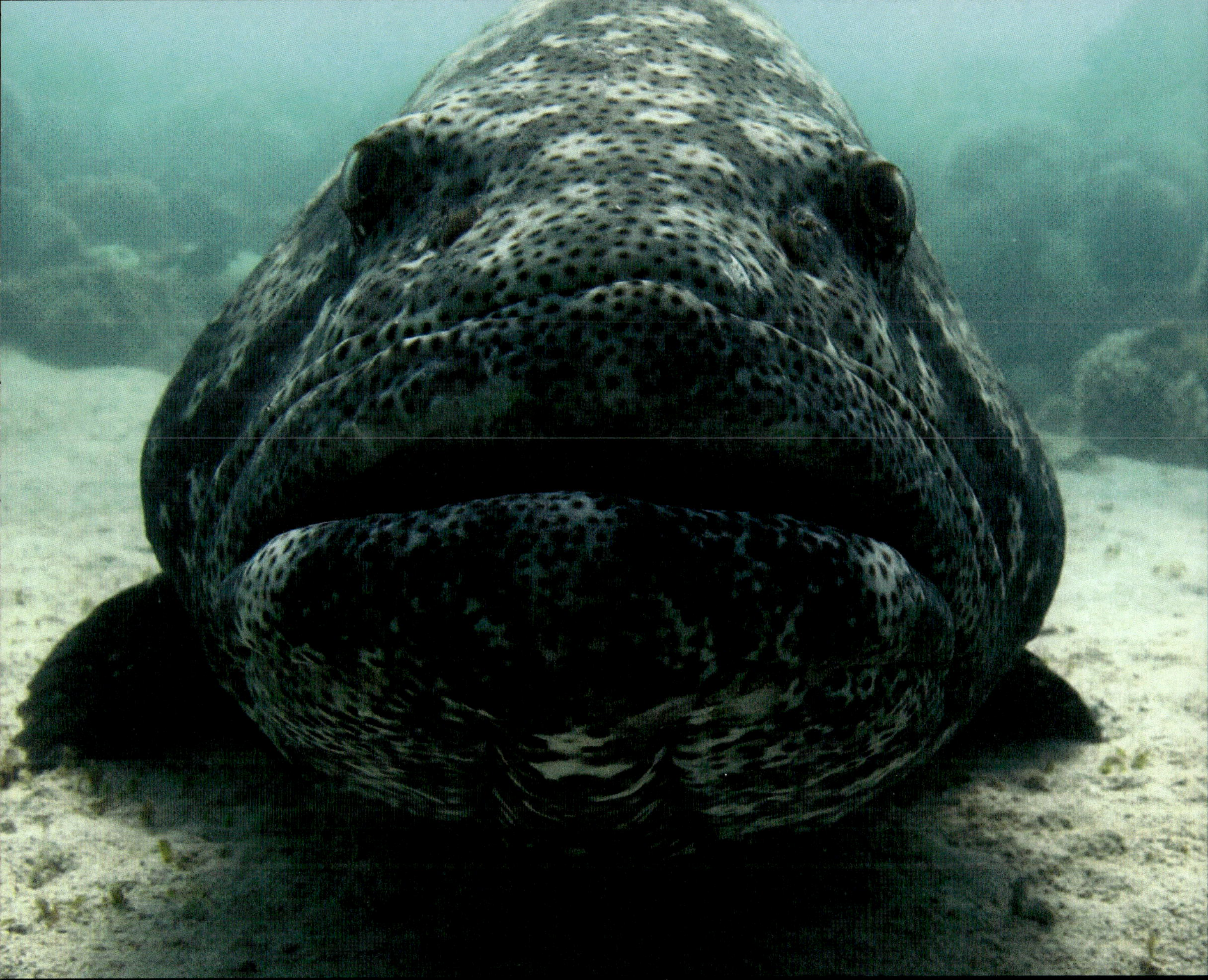

"He who knows others is wise.
He who knows himself is enlightened."
- Lao Tzu

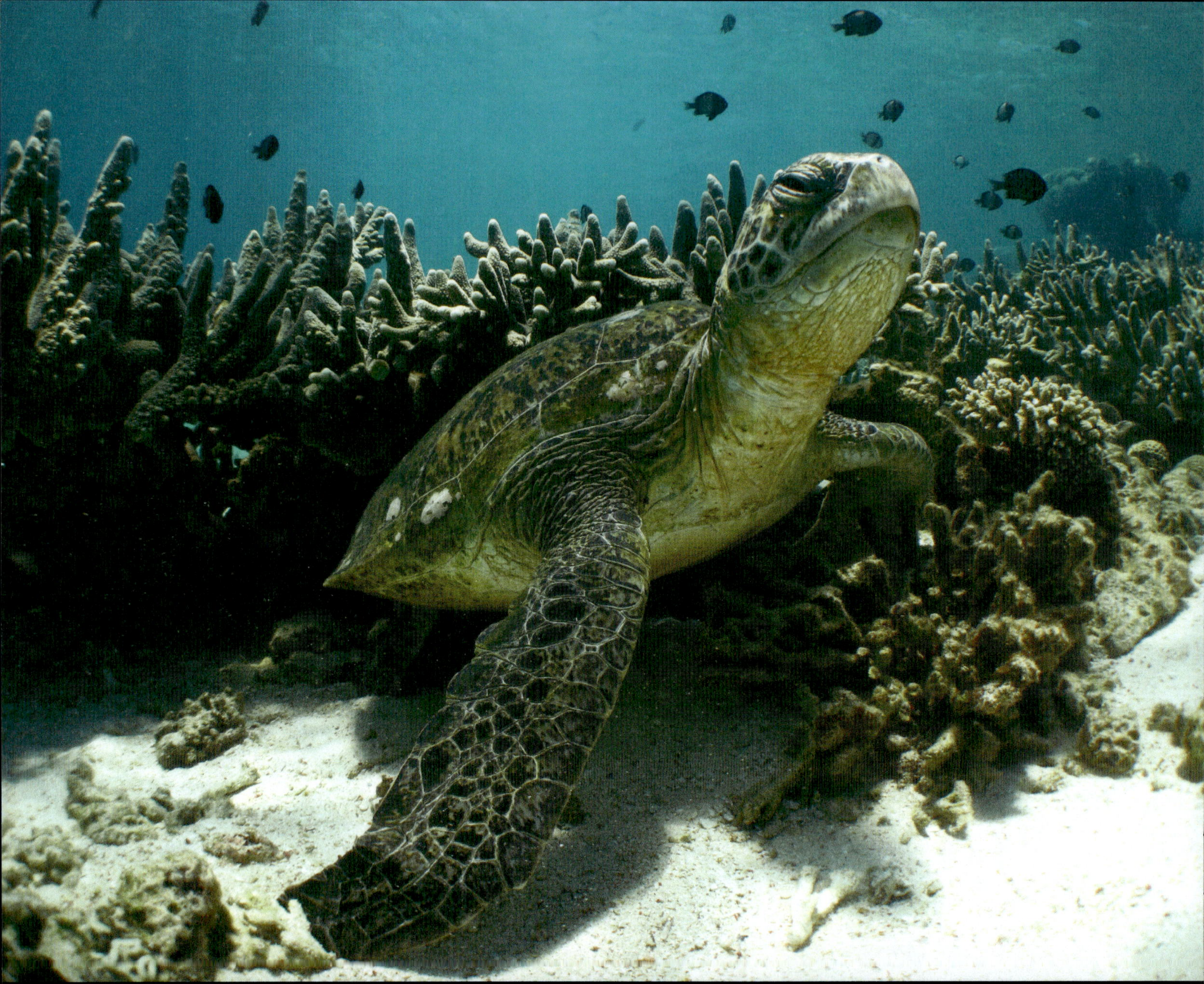

"Smile, breathe and go slowly."
- Thich Nhat Hanh

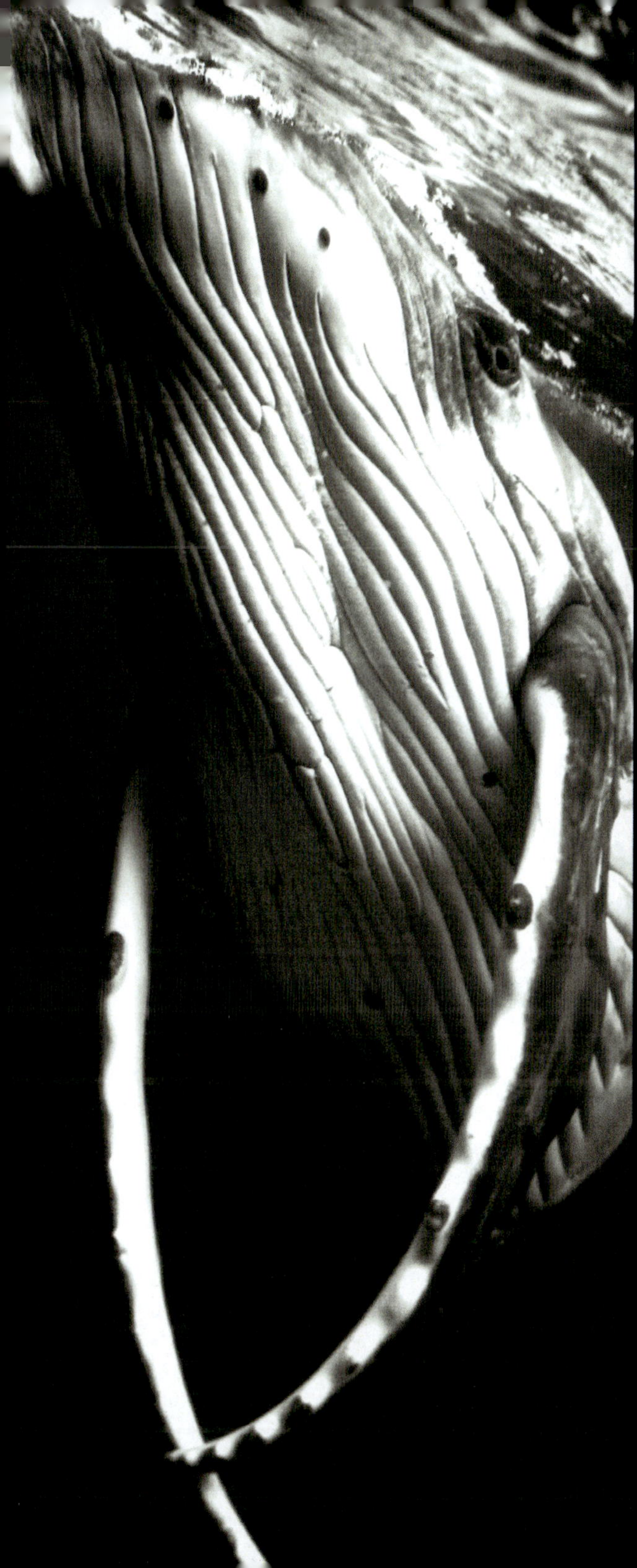

"In shallow men the fish of little thoughts cause much commotion. In oceanic minds the whales of inspiration make hardly a ruffle."
- Hindu wisdom

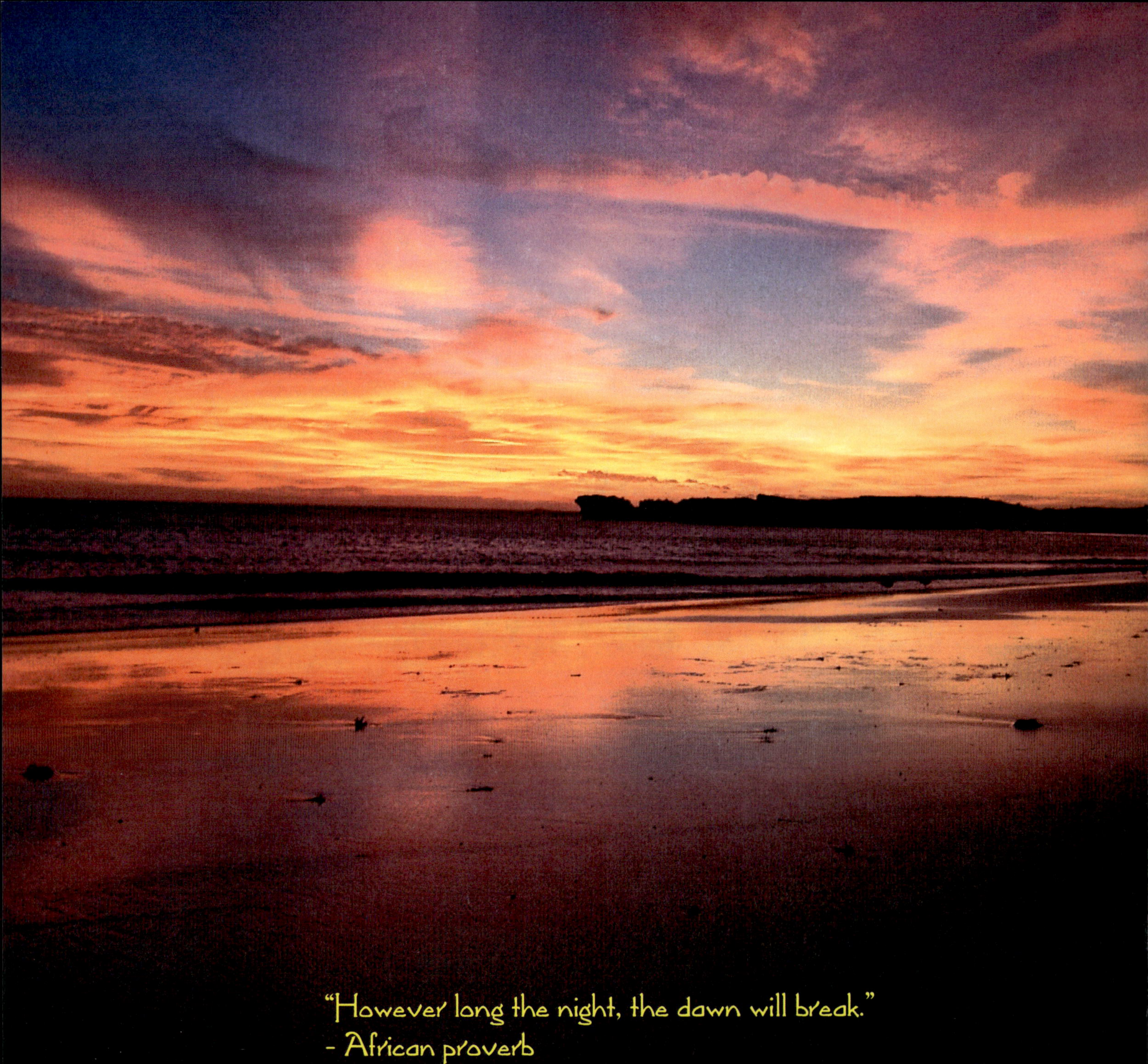

"However long the night, the dawn will break."
- African proverb

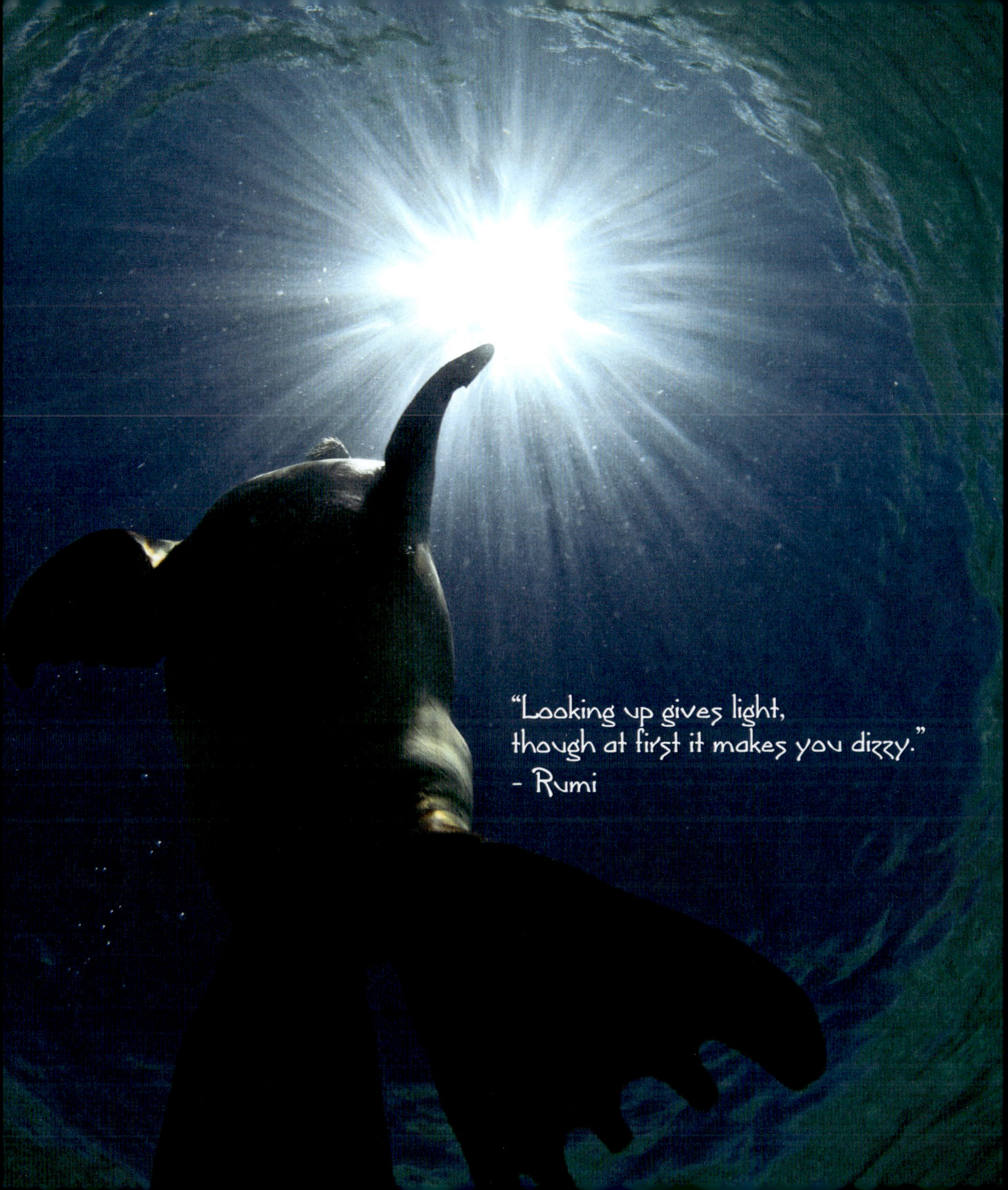
"Looking up gives light,
though at first it makes you dizzy."
- Rumi

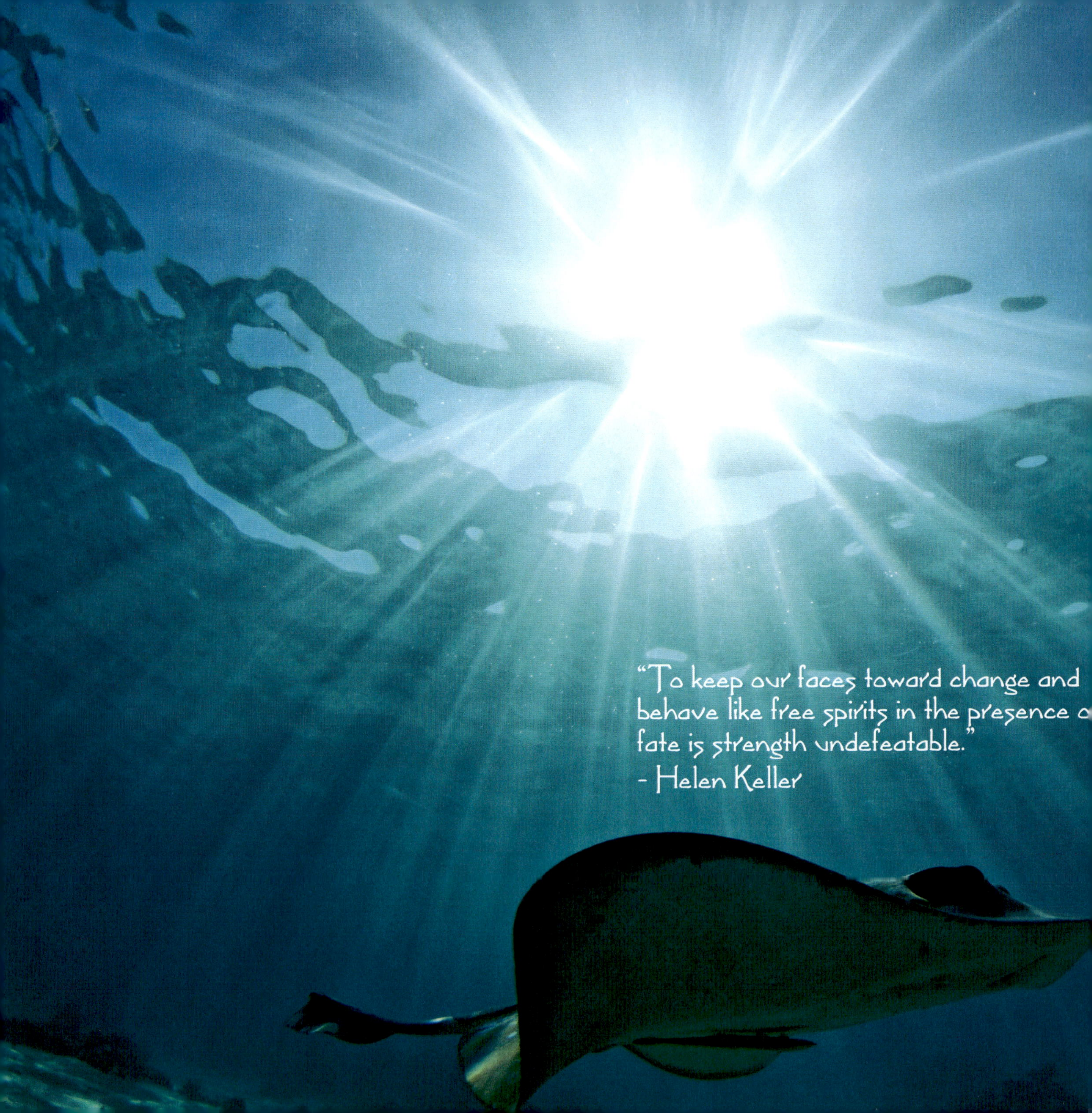
"To keep our faces toward change and
behave like free spirits in the presence o
fate is strength undefeatable."
- Helen Keller

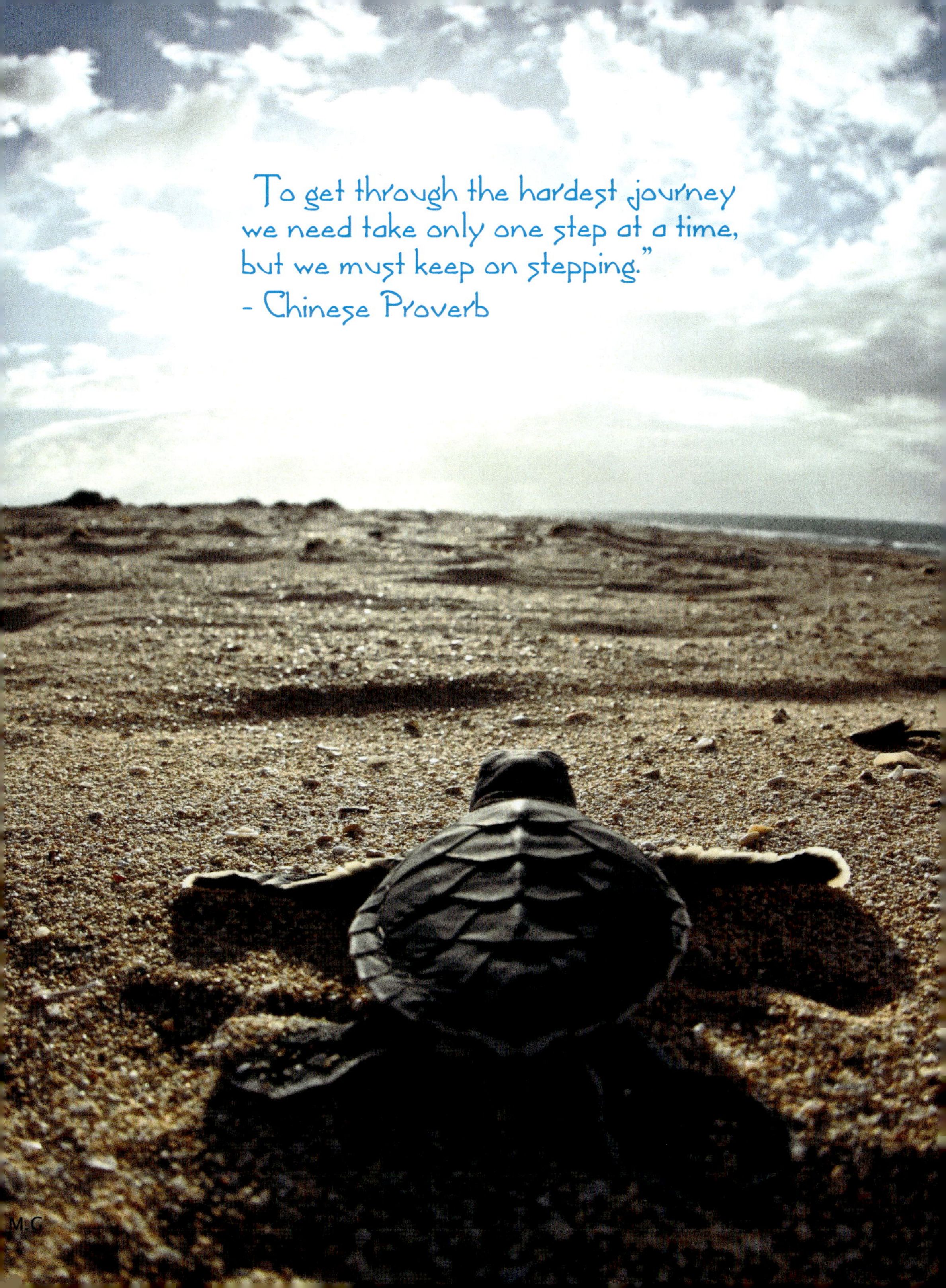
To get through the hardest journey we need take only one step at a time, but we must keep on stepping."
- Chinese Proverb
M.G

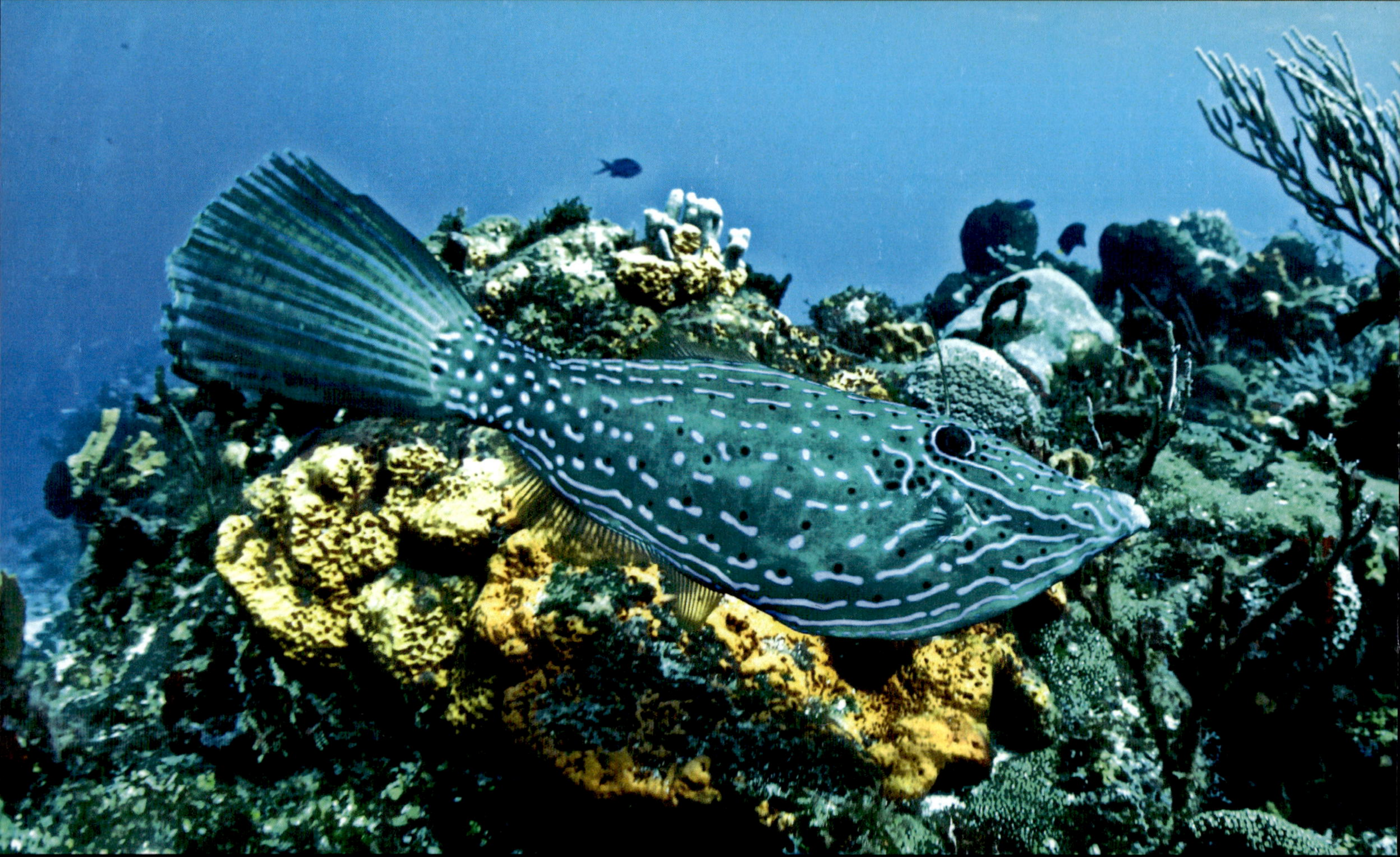

"Better to be without logic
than without feeling."
– Charlotte Bronte

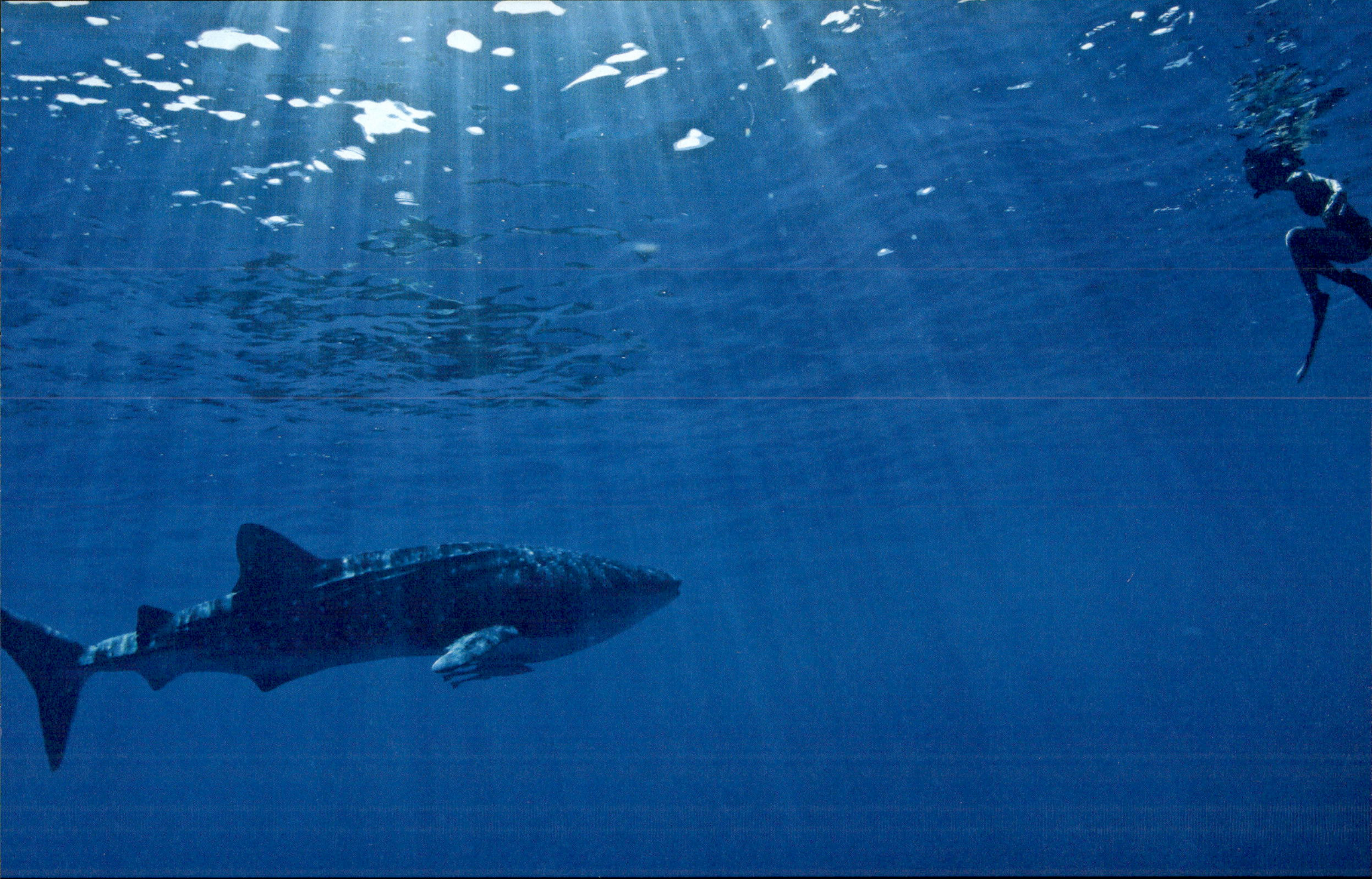

"How strange that Nature does not knock,
and yet does not intrude!"

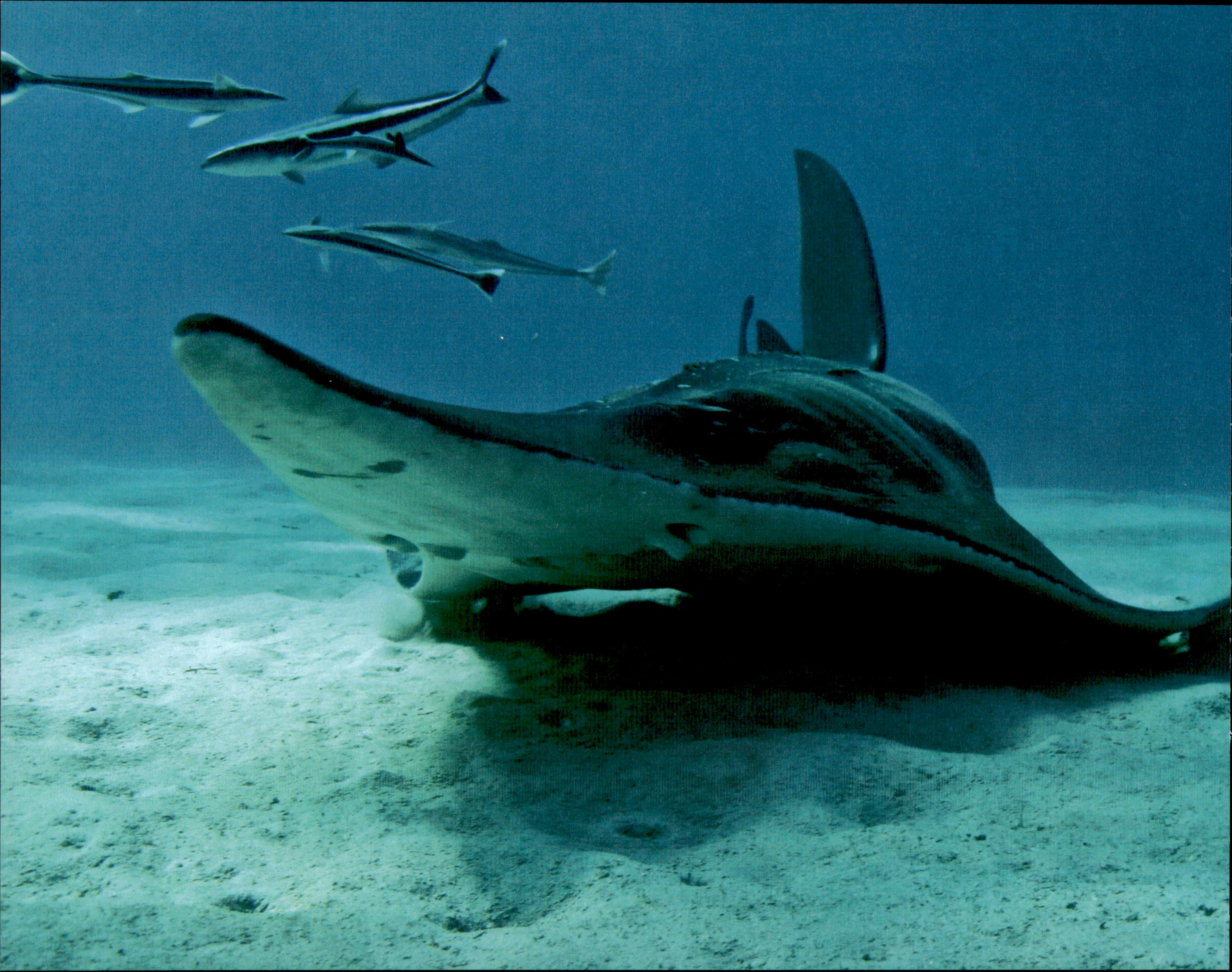

"Everything has beauty, but not everyone sees it."
- Confucius

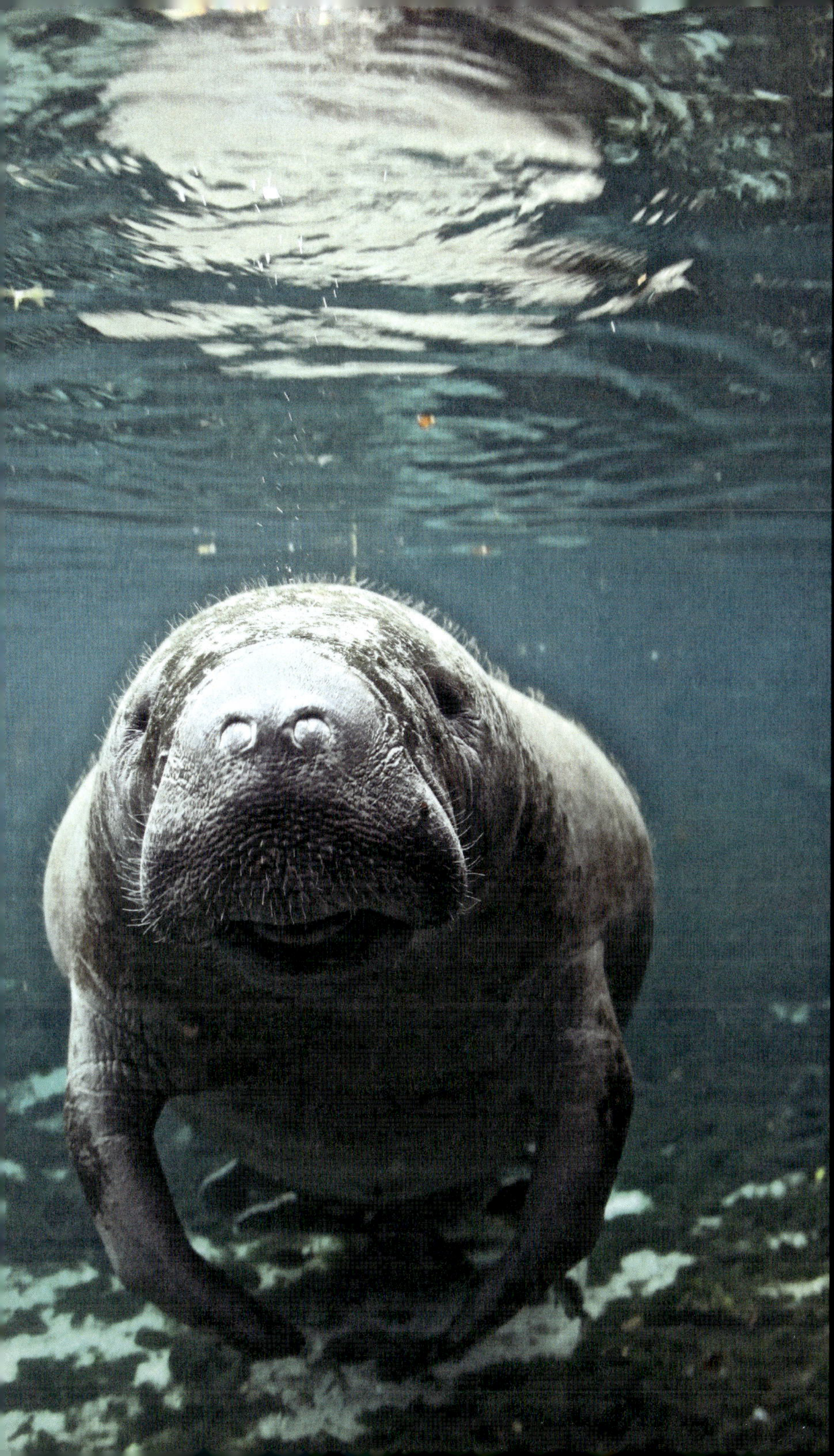

"My riches consist not in
the extent of my possessions,
but in the fewness of my wants."
- Joseph Brotherton

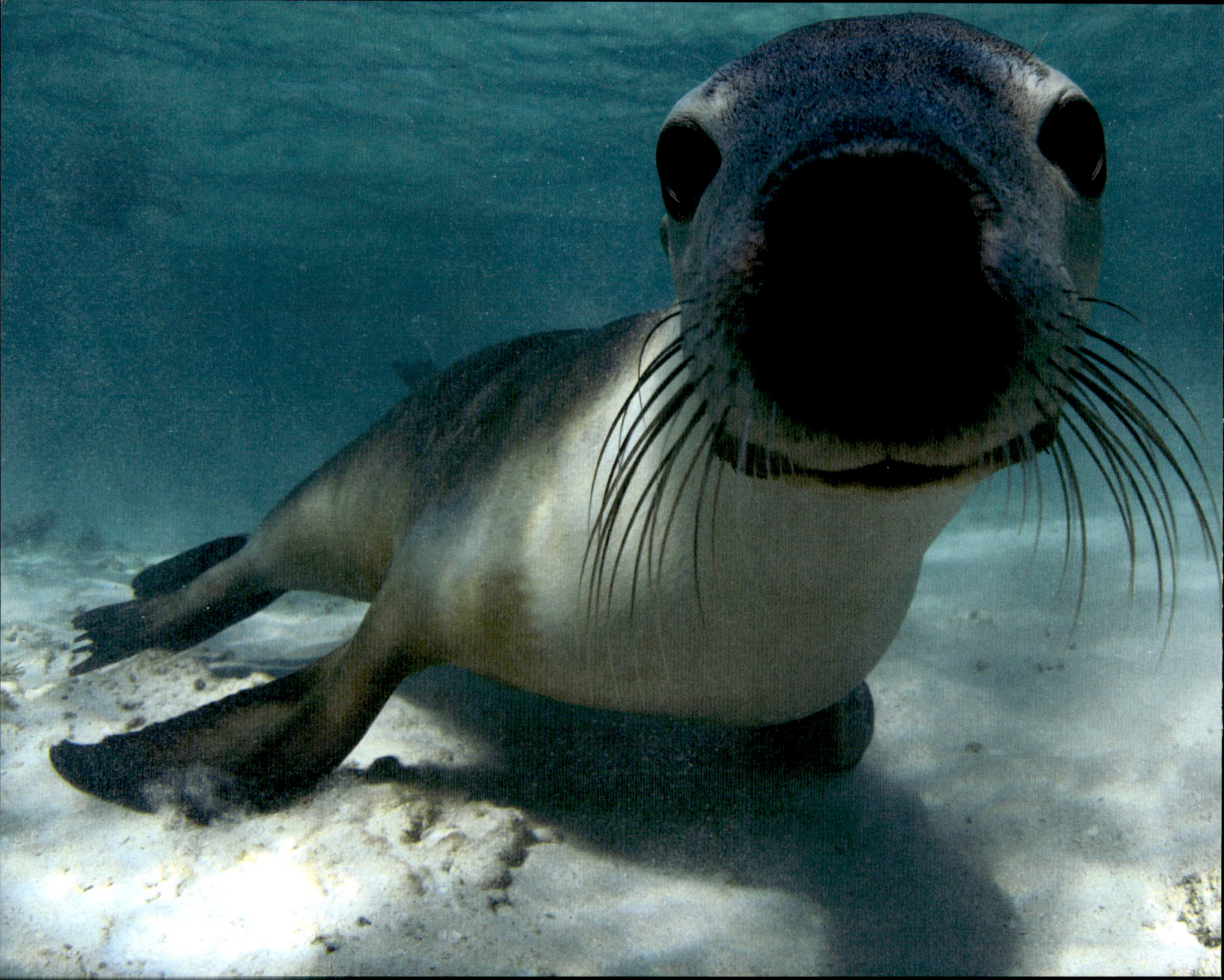

"A loving heart is the truest wisdom."

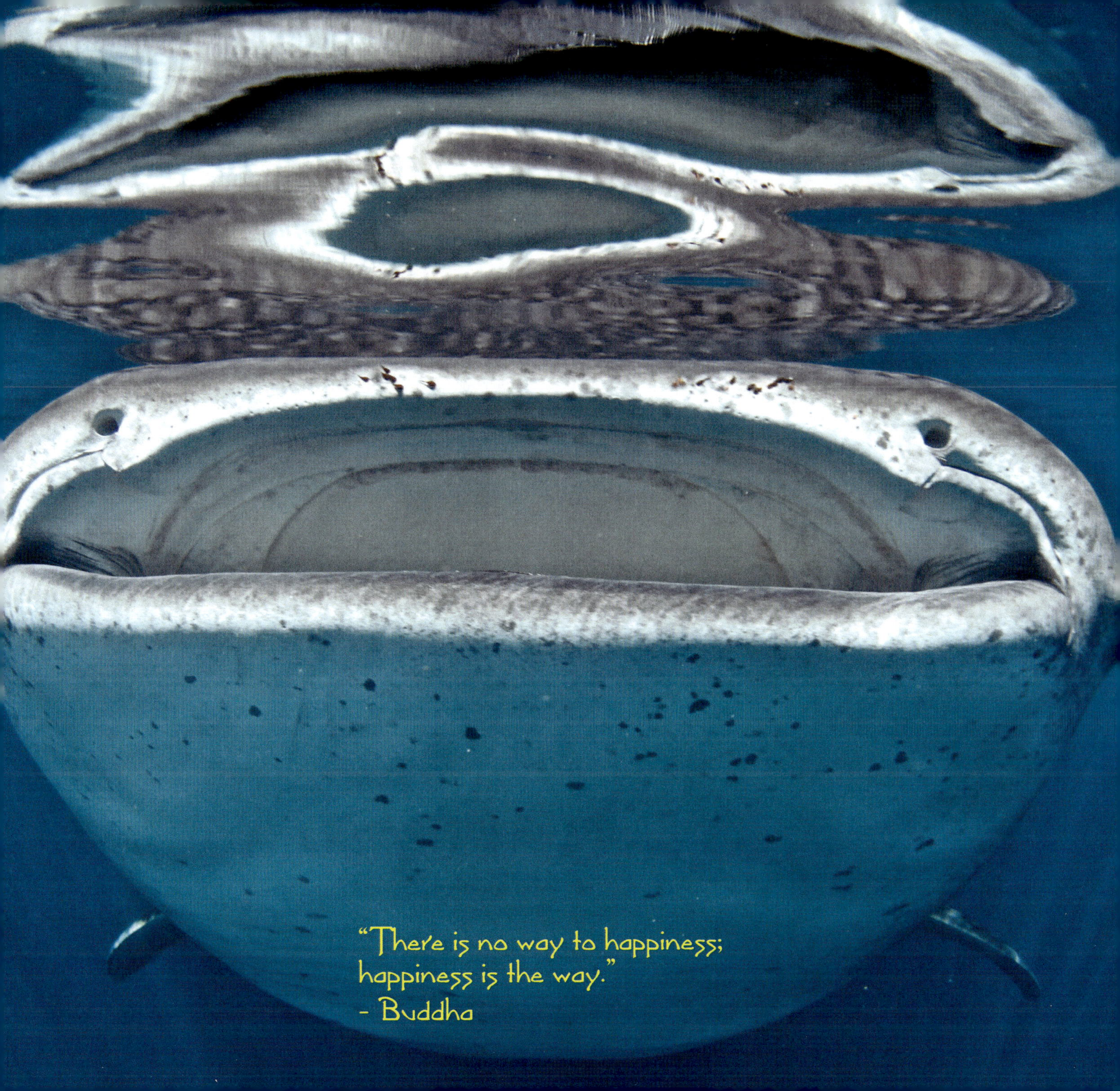

"There is no way to happiness;
happiness is the way."
- Buddha

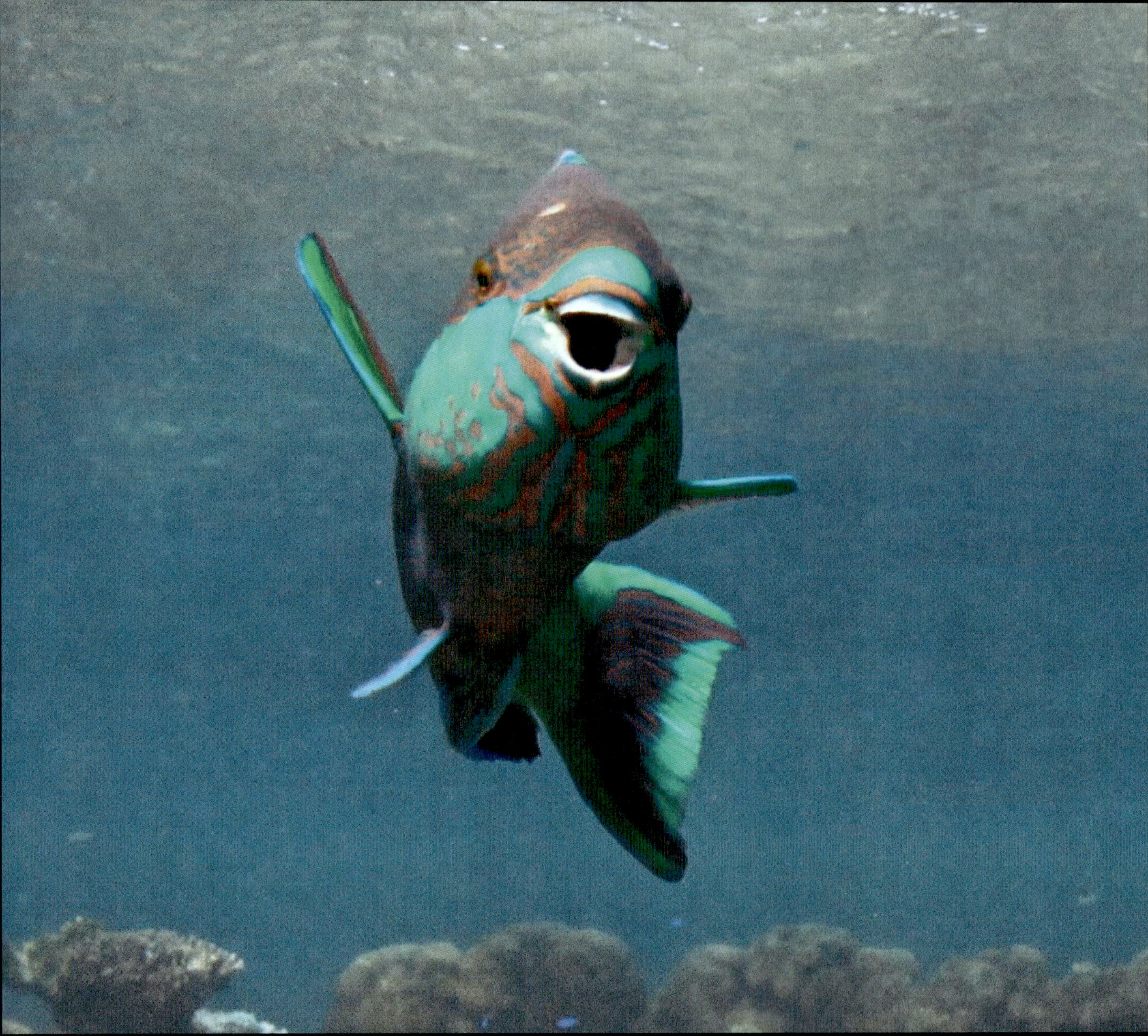

"You have succeeded in life when all you really want
is only what you really need."
- Vernon Howard

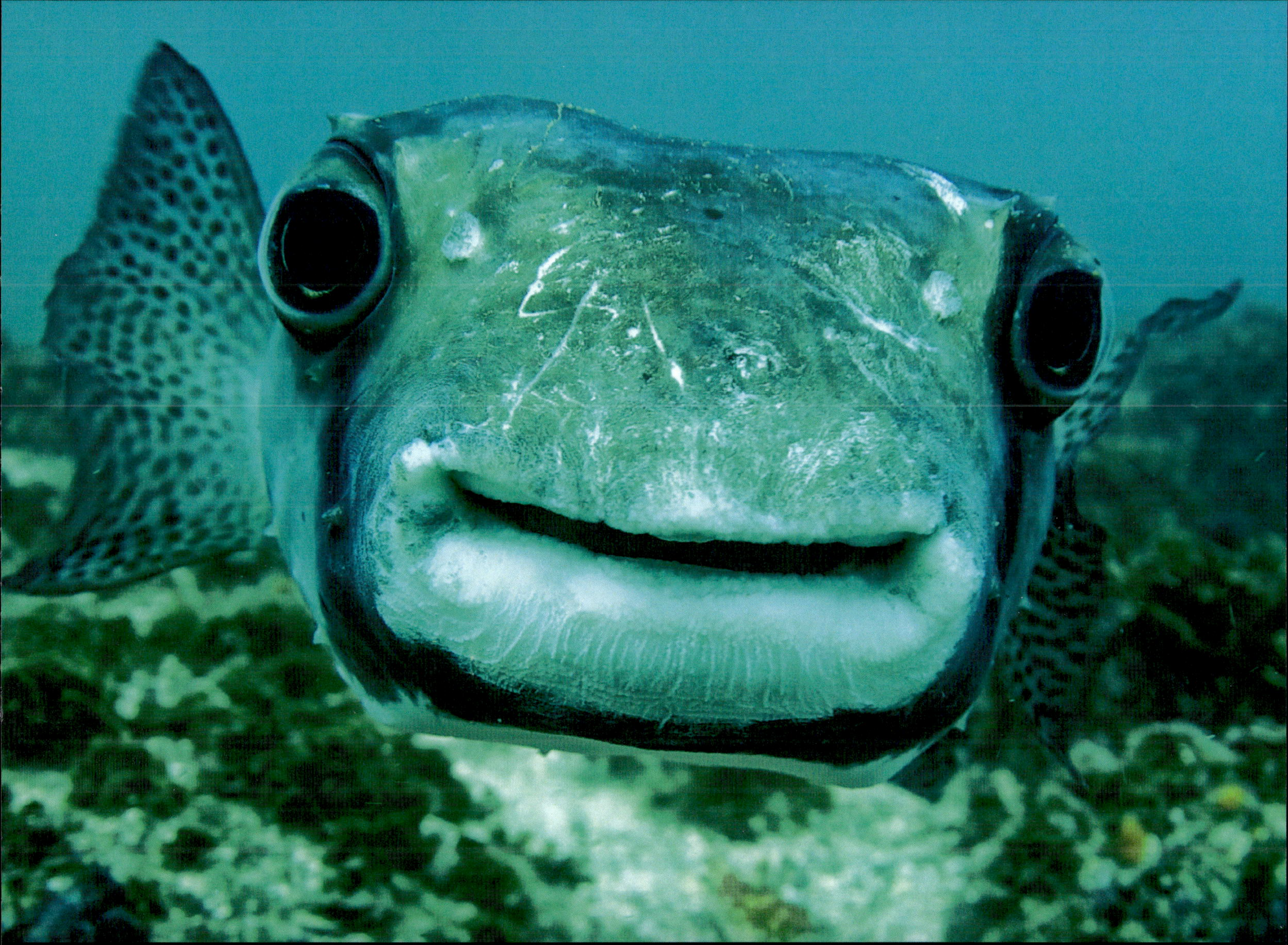

"Perfection has one grave defect:
It is apt to be dull."
- William Somerset Maugham

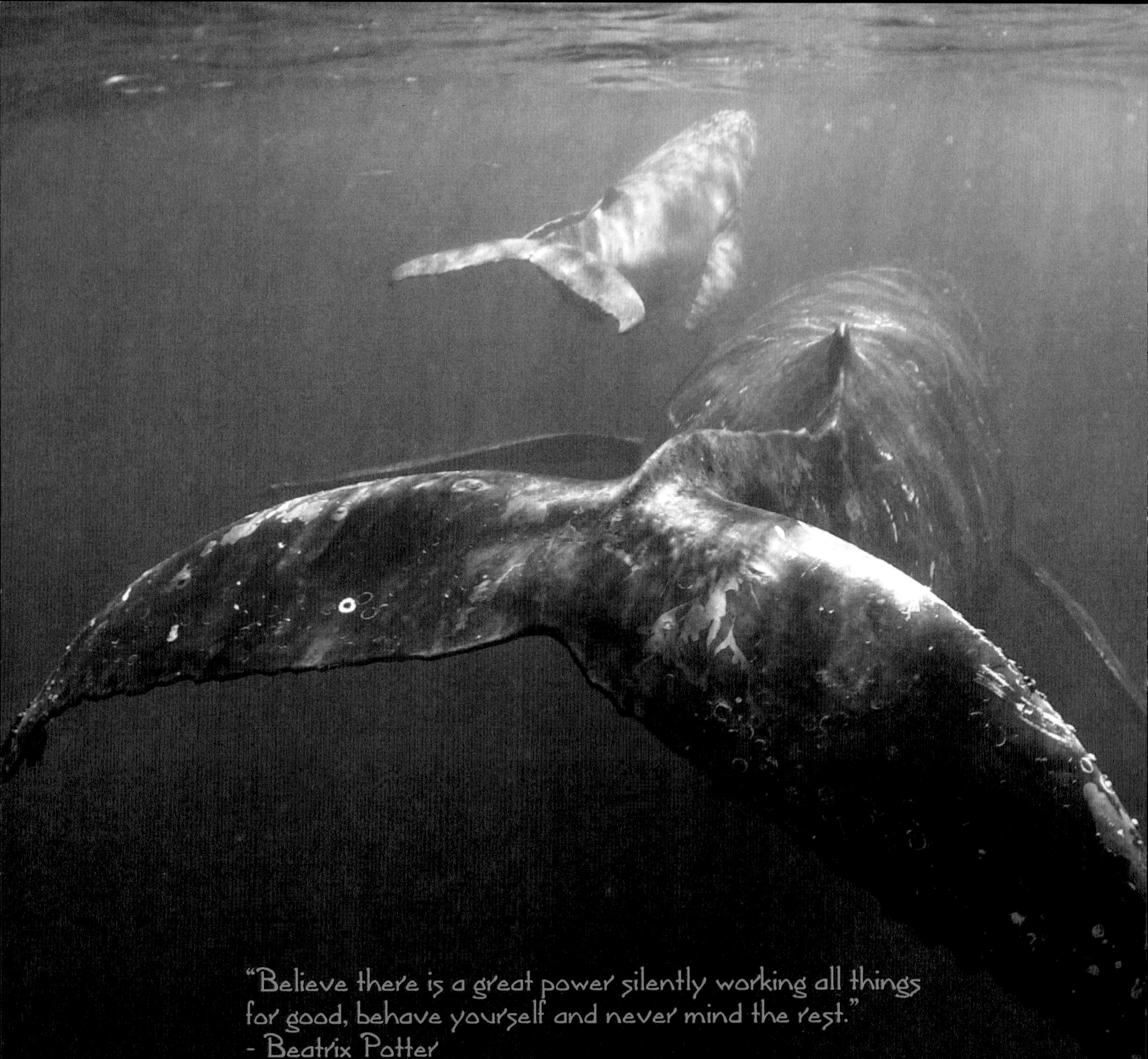
"Believe there is a great power silently working all things
for good, behave yourself and never mind the rest."
- Beatrix Potter

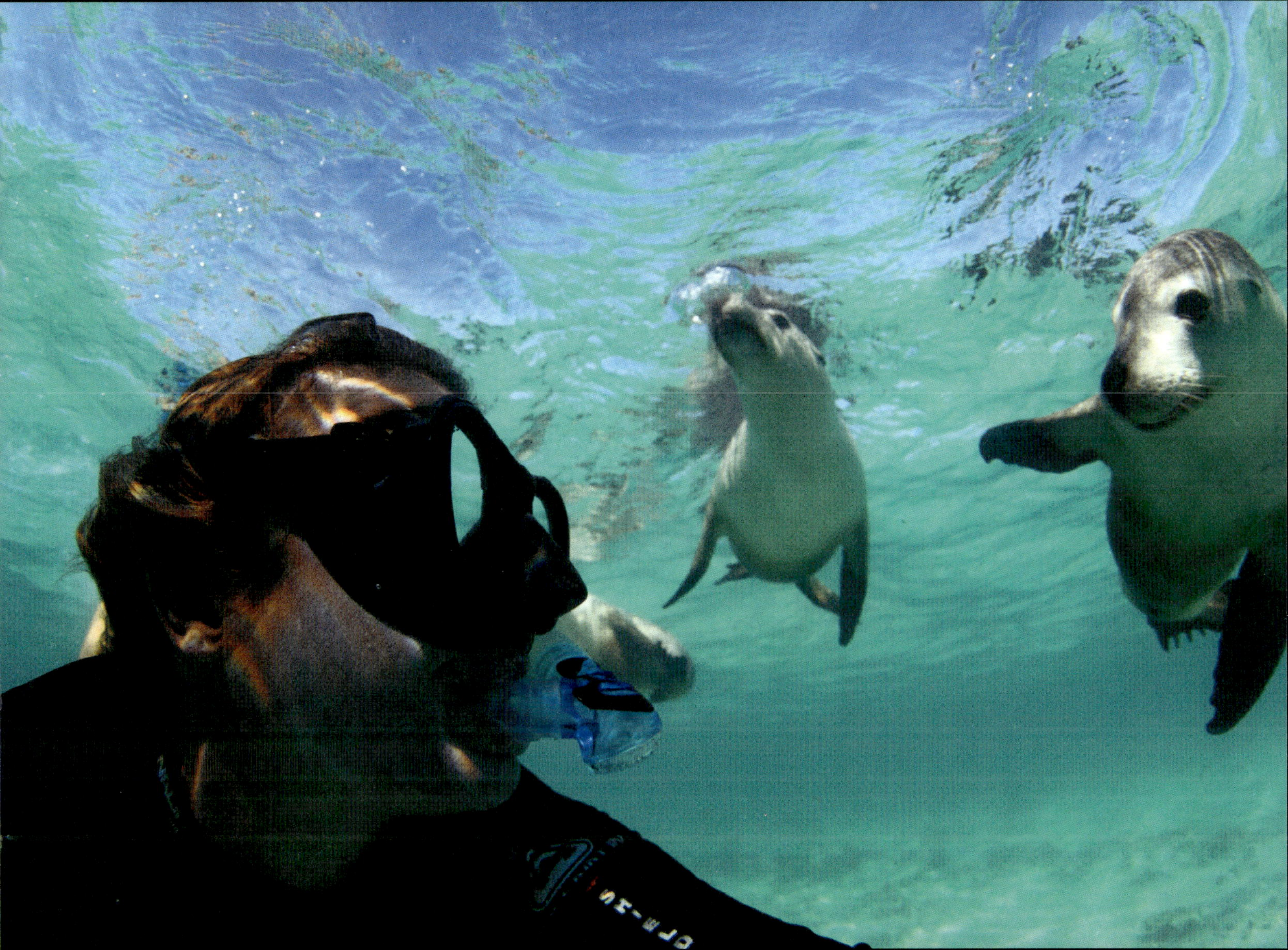

"Time you enjoy wasting is not wasted time."
- Marthe Troly-Curtin

Glass House Books
an imprint of Interactive Publications
Treetop Studio • 9 Kuhler Court
Carindale, Queensland, Australia 4152
sales@ipoz.biz
http://ipoz.biz/

First published by Glass House Books, 2017

Printed in 14 pt Argonaut.

National Library of Australia Cataloguing-in-Publication data:

Creator:	Green, Margaret Elizabeth, 1981- photographer.
Title:	Mermaids and monsters / photographer : Meg Green.
ISBN:	9781925231465 (hardback)
Notes:	Includes bibliographical references and index.
Subjects:	Marine animals--Australia--Pictorial works.
	Marine biology--Australia--Pictorial works.
	Ocean--Pictorial works.

Images by Meg Green
Book design by David P Reiter.

INDEX OF IMAGES

CREDITS

Page 1: Nigro., Samuel A., (2013). *The Soul of the Earth: Condensed version of Everybody for Everybody*. Xlibris Corporation.

Page 2: Feigenbaum, Edward. (18 Feb. 1989). *The Rise of the Expert Company. New Scientist Magazine*. Vol 121, No. 1652.

Page 3: The Southern Baptist Radio and Television Commission, (1972). *Home*, written by Ted Perry, inspired by a 19th century speech of Chief Seattle.

Page 4: Albert Einstein - interview with G. S. Viereck (26 October 1929).

Page 5: Homsett, Michael C. and Thomsett, Linda Rose (2009). *A Speaker's treasury of quotations: maxims, witticisms and quips for speeches and presentations*. McFarland.

Page 6: Demakis, Joseph. *The Ultimate Book of Quotations*. Lulu.com.

Page 7: Kirov, Blago. (2016). *Henry Ford: Quotes & Facts*. Blago Kirov

Page 8: Stone, Joshua David. (2001). *The Full Spectrum Synthesis Bible: Wisdom Quotes of the Masters of Religious and Spiritual Paths*. Universe.

Page 9: Bolsta, Phil. (2014). *Through God's Eyes: finding peace and purpose in a Troubled World*. Phil Bolsta.

Page 10: Chan, Wing-Tsit. (2008). *A source Book in Chinese Philosophy*. Princeton University Press.

Page 11: Leighton, Roy., Kilbey, Emma., Bill, Kristina (2011). *101 Days to Make a Change*. Crown House Publishing.

Page 12: Bolsta, Phil. (2014). *Through God's Eyes: finding peace and purpose in a Troubled World*. Phil Bolsta.

Page 13: Ibid.

Page 14: Helminski, Camille and Kabir. (1999). *Rumi Daylight: A Daybook of Spiritual Guidance*. Shambhala.

Page 15: Tatham, J. (2005). *Heart Lessons: A Novel about the Spiritual Journey*. iUniversity.

Page 16: Reiter, Laura (2014) *Beginner's Guide to Abstract Art: Making abstract art in watercolour, acrylics, mixed media and collage*. Pavilion Books.

Page 17: Bronte, Charlotte (2010). *The Professor* (Bibliolis Classics) Bibliolis Books.

Page 18: Bennett, Paula. (1990). *Emily Dickinson: Woman Poet*. University of Iowa Press.

Page 19: Bolsta, Phil. (2014). *Through God's Eyes: finding peace and purpose in a Troubled World*. Phil Bolsta.

Page 20: Ibid.

Page 21: Ibid.

Page 22: Ibid.

Page 23: Rick Reed (2016) *Good stuff for your heart and mind – a book of quotes*. 2nd edition. Lulu.com.

Page 24: Thomsett, Michael C., Thomsett, Linda Rose (2009). *A Speaker's Treasury of Quotations: Maxims, Witticisms and Quips for Speeches and Presentations*. McFarland.

Page 25: Potter, Beatrix (2012). *The Journal of Beatrix Potter from 1881 to 1897*. Penguin UK.

Page 26: Troly-Curtin, Marthe (1911). *Phrynette*, J.B. Lippincott Company, Philadelphia and London.

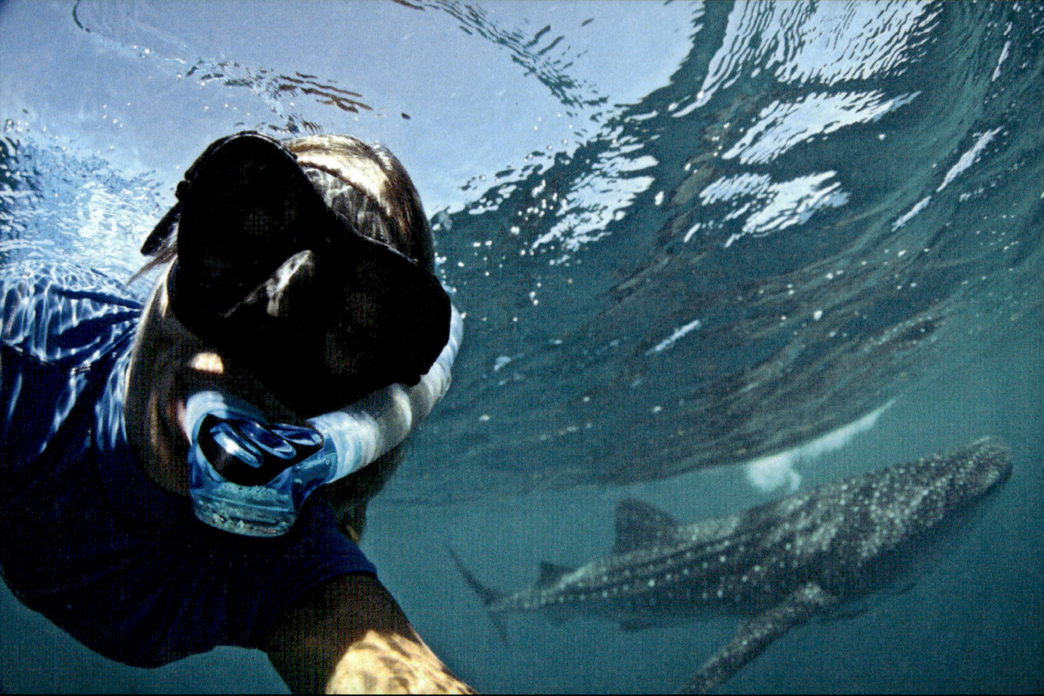

Meg Green is a marine biologist, photographer and educator who has participated in scientific, photographic and education projects the world over. An advocate of "living the dream", when Meg is not entertaining youth with fantastic "fishy" facts or photographing undersea delights, she spends time exploring and enjoying the wild places of Australia from the comfort of her mobile home "Gus". She currently resides in far north Queensland, Australia, and this is her first publication.

Check out more of Meg's photographs on Instagram: http://tinyurl.com/m76tles

Made in United States
Orlando, FL
06 November 2026

90230156R00021